AIR PO
AND ARMIES

AIR POWER AND ARMIES

J. C. Slessor,
Wing Commander

Foreword by
Phillip S. Meilinger

The University of Alabama Press
Tuscaloosa, Alabama

∞

The paper on which this book is printed meets the minimum requirements of
American National Standard for Information Sciences-Permanence of Paper for
Printed Library Materials, ANSI Z39.48-1984.

Library of Congress Cataloging-in-Publication Data

Slessor, John Cotesworth, Sir, 1897–
 Air power and armies / J.C. Slessor ; foreword by Phillip S. Meilinger.
   p. cm.
 Originally published: London : Oxford University Press, 1936.
 Includes bibliographical references and index.
 ISBN 978-0-8173-5610-1 (pbk. : alk. paper) — ISBN 978-0-8173-8330-5
(electronic) 1. Air power—Great Britain—History. 2. World War, 1914–
1918—Aerial operations, British. 3. Aeronautics, Military—Great Britain. I.
Title.
 UG635.G7S45 2009
 358.4'03—dc22

                                                              2009030308

To the memory of
LIEUTENANT ANTHONY BRISTOW SLESSOR
52ND LIGHT INFANTRY
Died Mandalay 19. xii. 32

FOREWORD

John Slessor was one of the Royal Air Force's most brilliant thinkers, while also being a successful combat commander. He led Coastal Command to victory in the Battle of the Atlantic during World War II. Later, Slessor was named chief of the air staff and retired from that position in 1953 as a marshal of the R.A.F.

He was a prolific writer. In addition to two volumes of memoirs, he wrote three books after retirement that dealt with NATO strategy and the role of airpower in the defense of Europe.[1] His most important work, however, was written between the world wars and concerned the employment of airpower in conjunction with a land campaign—*Air Power and Armies*. In emphasizing this type of air operation, Slessor was unique among his R.A.F. colleagues, who generally wrote either about strategic bombing or of narrow, tactical concerns. Although not discounting the importance of such subjects, Slessor concentrated instead on how best to use airpower as a complement to a field army. In essence, he stressed the need for air superiority, followed by systematic and relentless attacks against the enemy army's lines of supply—what today is termed air interdiction.

Slessor had fought in the R.A.F. during the Great War, been wounded in action, and served with distinction. Afterward, he worked on the staff of Air Chief Marshal Sir Hugh Trenchard. From his chief, Slessor learned of strategic bombing and became an advocate, but he also was a tactical air expert and had commanded an air support squadron in the 1920s. As a result, in 1931 Trenchard selected Slessor to become the R.A.F.'s instructor at the Army Staff College at Camberley. This tour solidified his knowledge and understanding of ground operations, and when combined with his combat experience as well as the tenets of strategic airpower learned at Trenchard's knee, Slessor became an unusually broad and flexible thinker. His challenge was to teach army officers the various roles and capabilities of airpower, while emphasizing the area of most concern to them—tactical cooperation. This required a delicate balance, but one he achieved nicely.

> writing to an Army audience

At the end of 1934 Slessor left Camberley for a tour in India. While there, he took the lectures he had given at the Staff College, refined them based on his current flying experiences in the Waziristan Campaign, and in 1936 published *Air Power and Armies*. In my view, this was the best treatise on airpower theory written in English before World War II.

Although acknowledging lessons from the Great War, Slessor believed that the character of war had dramatically changed since then. The evolution of the tank and airplane meant that the trench stalemate of the Western Front would be an aberration. In the future, war would be dominated by highly maneuverable armies. He believed airpower would play a dominant role in this new war.

As a result of the book's focus on an assumed land campaign, Slessor stated that airpower's role was to help defeat the enemy's army and air forces. This meant attacking the communications and supply lines of those forces—interdicting them—a subject generally overlooked by other air theorists. The first requirement for such air action, however, was to obtain air superiority. Slessor was certain that ground operations would not be successful without it. Significantly, he argued that air forces could conduct an air superiority campaign while simultaneously carrying out strategic strikes, cooperating with the army at the tactical level, and flying interdiction sorties. This ability to conduct parallel and not merely sequential combat operations was one factor differentiating airpower from surface forces.

In *Air Power and Armies*, Slessor focused on the theater—what today is termed the operational level of war—arguing that the neutralization of key nodes at that level would prevent effective military operations. He believed the enemy should be attacked, repeatedly, as far from the battlefield as possible. Airpower could seal off the enemy's forces and strangle them into submission. Interdiction should thus be the primary air mission when cooperating in a land campaign. In this regard he tended to favor supply interdiction (material and equipment) over force interdiction (troops and combat vehicles). Maintaining movement by rail and road was virtually impossible in daylight for the side that had lost air superiority. Cutting off all supplies and communications was not likely, but they could be severely curtailed and the enemy forces thus paralyzed. He elaborated on how to affect such paralysis.

Slessor believed that the new machinery of land warfare required ever increasing amounts of supplies, especially petroleum products and ammunition. This would in turn demand increased reliance on rail transport. Here was an Achilles' heel. In Slessor's view, railroads, especially the marshaling yards, were highly vulnerable to air attack and, therefore, were a growing liability to the side that lost air superiority. Of note, he pointedly warned his army colleagues that they must take greater care to protect their own supply lines from enemy air attacks. Friendly air superiority should not be assumed.

Significantly, Slessor recognized that effective interdiction required cooperation between air and ground units, reasoning that there were occasions when ground forces should support the air effort—a heretical belief among most ground officers at the time. Finally, he contended that airpower must be commanded and directed by an airman who was equal in authority to the ground commander. These two individuals and their staffs were to collaborate in the design and implementation of the theater commander's overall plan.

Air Power and Armies is an outstanding book with several notable aspects: its emphasis on the need for specialized air intelligence, and the detailed discussions of air superiority and of an army's center of gravity—its supply lines. Slessor argued that when cooperating in a land campaign, airpower was wasted if used merely as a tactical weapon; rather, airpower should concentrate on the disruption, destruction, and neutralization of enemy armaments and supplies— interdiction. It was an idea proven accurate in World War II.

Phillip S. Meilinger

1 Slessor's memoirs are *The Central Blue: Recollections and Reflections* (London: Cassell, 1956) and *These Remain: A Personal Anthology* (London: Michael Joseph, 1969). His works on defense policy include *Strategy for the West* (London: Cassell, 1954), *The Great Deterrent* (London: Cassell, 1957), and *What Price Coexistence?* (London: Cassell, 1962).

INTRODUCTION

THIS book is based on a series of lectures delivered at the Staff College at Camberley between 1931 and 1934. It deals with the action of the Royal Air Force in one special set of conditions, namely when the Empire is engaged in a war in which it has been necessary to send an Army and Air expeditionary force to fight in an overseas theatre of war. The first and most important commitment of the Royal Air Force is, of course, the defence of Great Britain against air attack; and intimately connected with this commitment is the provision of an air expeditionary force to co-operate with the army in a campaign overseas. And as long as we live in a world which maintains huge national forces numbering millions of men and consisting largely of the traditional arms—infantry, artillery, and cavalry, it is obviously important that all officers, at least of the army and the air force, should understand how the new power of the air is likely to affect the problems of land warfare.

It may be as well to anticipate two criticisms which may reasonably be directed against this book, both of which may arise from the fact that I have drawn largely on the recorded experience of the last war in order to illustrate my points. The first is that many of the comments upon and criticisms of the conduct of air operations in the last war are based on 'wisdom after the event'. They are—quite frankly and deliberately so. But there is no question of blaming anybody for any sins of omission or commission; my sole object has been to draw conclusions on which to base useful lessons for the future. After all, the really important function of any kind of military history is not primarily to serve as interesting material for the general reader, but to enable commanders and staff officers of the future to be wise *before* the event, and to learn not only from the successes but from the failures of their predecessors. There is a great deal in the history of the War in the Air which may serve as a model for the future. But there was inevitably also a certain amount which might have been done better—inevitably because we were all of us amateurs at a new art; and there could be nothing more dangerous than to sit back and assume complacently that all that we did was good. This book, therefore, is written in no spirit of destructive criticism. I believe that on the whole we had in the last war the best led, best trained, and most efficient air force in the world—with our late enemies the Germans a very good second. Since then we have made great progress in the art of air warfare. The technical efficiency of our aircraft is to-day extremely high, and beyond our wildest hopes in 1918. Our training, and in particular our weapon-training, is of a very high order indeed; and we have to-day in our service manuals and training establishments the fruits of years of study and discussion on the strategy and tactics of air warfare. In the last war our commanders and staff officers had none of these advantages. The great air forces of which we then disposed were a mushroom growth; and the very rapidity of that growth, allied to their relative technical inadequacy, and the natural bent of minds to whom the problems and potentialities of air warfare were entirely new, had the inevitable result that they were not always used to what we should to-day consider the best advantage. Perhaps partly for this reason there is a tendency to forget that our only practical knowledge of air operations in first-class warfare is based on the experience in many theatres between 1914 and 1918; and hence to ignore the many valuable lessons—some of them of a negative order—which emerge from a study of those campaigns.

From this may arise the second criticism which it is desired to anticipate. It will undoubtedly be said that modern developments have altered cases; that the conditions of the last war are unlikely to be repeated; and that a close study of an operation such as that of the Amiens battle of August 1918, contained in Chapters VIII–X of this book, is a waste of time, because its essential characteristics are unlikely to be reproduced. It is obviously true that the sealed-pattern, trench-warfare, infantry and artillery battle on the 1914–18 model will never be seen again. Whether or not warfare on anything akin to traditional lines is altogether a thing of the past is a question to which an answer is suggested in the concluding chapter of this book. It is fashionable nowadays to represent the war of the future as being inevitably an affair of lethal gases and bacilli rained from the air exclusively upon the female and non-combatant sections of the populace in open towns. I would prefer to abstain from prophecy on that head—further than to suggest that it is dangerous to take for granted that military operations of any nature on the ground are, as yet, only a matter for the reminiscences of a modern generation of Old Kaspars.

In this connexion there is one point in particular which must be referred to, because it is a point of primary importance in British defensive policy.

'The reasons why England in the reigns of William and Anne felt constrained to take part in the wars against Louis, are the same reasons that have periodically guided her action in great European crises and colonial rivalry . . . the need to secure the safety of our small island by preventing the predominance of any one Power on the Continent—the Policy known as the Balance of Power; and the imperative demand on behalf of our maritime security that the Low Countries should not fall into the hands of a great military and naval Empire.'[1]

The Policy of the Balance of Power is—theoretically at any rate—out of date in these days of the League of Nations. But if the freedom of the Low Countries has been a cardinal point in our policy for reasons of maritime security in the past, it may be no less vital to us to-day for the added reason of air security. In air defence a first essential is depth, because depth means time and space—time in which to get warning and enable our fighters to reach their fighting height, and space in which to establish our bomber aerodromes well forward in the vicinity of the hostile air bases and vital centres. And although the rapidly increasing ranges of bombers will in the near future diminish the importance of this factor, it will still remain true that a much more intensive attack against this country could be sustained from bases on the Channel coast.

It is difficult, for obvious reasons, to be more explicit on this point. It must be sufficient to suggest that to ensure the integrity of the Low Countries, to prevent the establishment of hostile air bases within fatally close range of these coasts, military operations on the ground—though inevitably of a very different character from those of 1914-18—may again be necessary in the future as they have so often been in the past.

Therefore, since we must assume that military operations on the ground may take place again, of however different a nature,

---

[1] G. M. Trevelyan, *Blenheim*, p. 107.

it is obviously worth while trying to learn something from a study of the only first-class war in which aircraft have played an important part. If there is one attitude more dangerous than to assume that a future war will be just like the last one, it is to imagine that it will be so utterly different that we can afford to ignore all the lessons of the last one. The interval between the South African War and the Great War contains some interesting examples both of failures and achievements in this respect. For instance, in 1914 our system for the supply of artillery ammunition was based on the policy that it was required to maintain in the field a force of six divisions and one cavalry division 'during a campaign similar to that in South Africa'; the reserve stocks in hand and the system for replacement were based on the conclusions of a Committee which sat in 1904; and those responsible for our military policy did not revise their conclusions in the light of the lessons of the Russo-Japanese War, which was going on while that Committee were sitting. In other words, in 1914 they were making some of their preparations, not for the last war—that would have found them far better prepared than they were—but for the war before that, which was obviously unwise and dangerous, especially in view of the European situation and the war for which the army were almost openly training.

On the other side of the picture there is this to remember. The campaign of 1914 had very little in common with the South African War—though admittedly it was nothing like as different as the next land operations are likely to be from those of 1918. There is discernible nowadays a tendency to make more of the defects of the British Regular Army of 1914 than of its qualities—possibly on account of some admitted deficiencies in its command and direction; but no one can deny that its organization, its training—particularly weapon-training—and its equipment were excellent, and worthy of the splendid quality of its personnel. And the reason that this was so was very largely because the General Staff in the decade following South Africa had studied and remedied many of the defects in direction, organization, training, and equipment which had disclosed themselves in that campaign. So, however different the next war may be, we can probably learn something even on points of detail from the lessons of the last. But the real value of intelligent and critical study of recent campaigns is that it does give us a sound grasp of the main underlying principles. This is a platitude, of course, and so is the statement that the basic principles of war—which are merely the basic principles of common sense applied to war—do not change; but they are both of them profoundly true. There does arise occasionally the military genius, the born commander who instinctively does the right thing on every occasion as it occurs. But the ordinary man is much more likely to do the right thing if he really understands why he is doing it, and what will probably happen if he does something else; and the best basis for sound judgement is a knowledge of what has been done in the past, and with what results. We in the air force have only one war to draw on for such knowledge; and therefore if a study of the air operations in that war helps us to understand the broad principles such as concentration, security, and offensive action, it needs no further excuse.

So it is in this spirit that this book is written. Not in a vein of facile destructive criticism, but as a reasoned attempt to examine some possibilities of air action in co-operation with an army in the field, in the light of experience of the last Great War; accepting examples from that war as models for the future when they appear admirable; and when they do not, frankly suggesting how they might have been bettered had there been then at our disposal the experience and knowledge which, in however imperfect a form, we have to-day.

It should, however, be clearly understood that the conditions envisaged throughout are those of a campaign on land in which the primary problem at the time is the defeat of an enemy army in the field. Although the great principles remain the same, their application must vary widely with the conditions of the war. And in a war against a great naval Power at sea, or when the principal threat to the Empire at the time is the action of hostile air forces against this country or its possessions, the aim and objectives of the air forces of the Empire will not be the same as those described in this book.

It must be explained that anything in the way of historical research into the operations of the air force in the last war is severely handicapped by the lack of comprehensive documentary records. A very close and excellent liaison existed between all air force commanders and the General Staff of the formations they served; and by far the greater part of the arrangements for air co-operation was done by personal discussion and verbal instructions, usually unconfirmed in writing. This system, though no doubt it often worked very well at the time, makes things very difficult for the student or military historian after the War. I have been able to consult a few of the commanders and staff officers who were concerned, for instance, in the arrangements for the Amiens battle; but it is notoriously difficult years afterwards to remember the atmosphere in which decisions were made, or to grasp what was in the minds of the men on the spot at the time merely from written records, however full and comprehensive they may be. Subject to these inevitable limitations every effort has been made to ensure that the narrative—even if in some places necessarily incomplete—is substantially accurate on matters of fact. Where suggestions or criticisms are offered on matters of opinion—for which, of course, I bear the sole responsibility—it should be remembered that 'what's done we partly may compute, but know not what's resisted'. And the decision or course of action under review may have been dictated by strong and adequate reasons of policy, military expediency, or the personal factor, known only to the responsible authorities concerned at the time.

When considering the possible course of air operations in the future, and comparing it with the recorded history of air operations in the past, the reader must always keep in mind the immense advance in technique and material efficiency since the War. This is a subject which obviously does not lend itself to detailed discussion in a book of this sort. Since the end of the War aircraft have made a great advance in performance and offensive power; and the next few years will probably see a still greater improvement both in speed and carrying capacity. These improvements in design, together with the development of engines burning heavy oil—which is a more economical fuel than petrol—will undoubtedly mean a very great increase in bomber ranges in the near future, possibly by 100 per cent. or more.[1] This is obviously the most important development in its strategical implications to the security of Great Britain. But apart from this, to overlook the very great increase in such

[1] See Lecture by S.-Ldr. R. V. Goddard, *R.U.S.I. Journal*, August 1934.

factors as bombing accuracy and the efficiency of air armaments, of organization, and of training, since British aircraft were last in action on a serious scale, would be to get a very false picture of the potentialities of air warfare in modern conditions.

Finally, I must acknowledge most gratefully the invaluable assistance of the many officers of the army and air force, and of the Air Historical Branch, whose ideas I have borrowed, and whose criticism and advice have helped me so much.

J. C. S.

QUETTA,
March, 1935.

CONTENTS

PART I. AIR SUPERIORITY

months of Great War—lessons—damage due to hangars getting burnt out—modern independence of hangars—need for dispersal of aircraft on aerodromes—results of overcrowding of aerodromes—the French experience—need for dispersal of aerodromes with fewer aircraft on each—the question of control—bombing of aircraft depots—British experience in France—lessons—facilities for replacement of casualties in aircraft—the time factor, results of bombing just before an important operation.

The principle of concentration—two main classes of objective, fighting troops and supply—definitions—supply includes production—relation between attack on production and more intimate forms of co-operation in the field—the moral effect of air action on the 'non-combatant' as a background to this chapter.

The object of attack on production—definition of war industries—war industrial areas of most European countries within air range of frontiers—effects of attack on production—lesson of German attacks on England—methods of attack on production—Foch's appreciation of 1918.

The two opposing schools of thought about attack on production as opposed to intimate co-operation in the field—fundamental unsoundness of both—answer is mobility and concentration on whichever is most vital at the time—illustrated by history of formation of the Independent Force in 1918.

Early attacks on German war industries—first conception of 'long-term' strategic possibilities of such attacks—creation of the Royal Air Force in April 1918—Sir F. Sykes as C.A.S.—situation on the Western Front at the time, the German spring offensives—Proposals for formation and command of Inter-Allied Bombing force—Trenchard assumes command of new Independent Force 6.vi.18—Question of Inter-Allied Bombing force decided—Foch instructed to draw up programme—Trenchard's views on the conditions in which bombing of Germany should be undertaken.

Contemporary events at G.H.Q.—the recommendations of the Transportation experts as to bombing of railways—Air Head-quarters objections on the score that insufficient aircraft were available to carry them out—the 'concentrated bombing' schemes—conclusion that employment of Independent Force against objectives in Germany during critical situation on Western front was an unjustifiable diversion—possibility of an Inter-Allied Air Reserve under Trenchard to be used at the decisive point.

The right time for attack on production—cannot entirely replace closer co-operation in land campaign—but will influence and should limit operations on the ground—attack on production in its wider sense. Fighting Troops and Supply (including production) not the only possible objectives even in land campaign—others may assume temporarily more vital importance—the German submarines in spring of 1917—enemy air forces in another war—air action must be concentrated against whatever objective is strategically most decisive at the time—importance of mobility.

The process by which the air plan is evolved—the three essentials—Intelligence, more detail, and greater depth of zone covered—Technical advice—Reconnaissance.

FIGHTING TROOPS as objectives—general rules—the aeroplane not a battle-field weapon—August 8th and Cambrai—movements of troops by road—armoured forces less vulnerable but more susceptible to interference with supply—need for depth for initial attacks on columns on roads—troop movement by rail—entrain-

CONTENTS

ment rather than detrainment the objective—mounted troops and animal transport very vulnerable—troops in rest or reserve—effects of permanent insecurity—the Étaples raid—Head-quarters, and cable communications—Megiddo.

LOW-FLYING ACTION—definition—assault aircraft not a special class in British service—high degree of air superiority essential—heavy casualties involved —*in attack* may be used temporarily to help break crust—but after initial break-in must be lifted against rear communications—pursuit the supreme opportunity —Megiddo—the Macedonian front—Vittorio Veneto—*in defence* modern developments may reduce value—rearguard actions during the retreat of the Fifth Army—the *control* of assault squadrons—definite initial orders essential—assault action after the first attack equally must be controlled—difficulty of getting information—the Flesquières incident—Report by O.C. 22 Wing after August 8th—importance of assault pilots knowing the ground—the German Staff paper on the subject.

SUPPLY as an objective—actually during battle attack on supply and fighting troops identical—importance during period of preparation—vast supply requirements of modern armies—Colonel Wingfield's article on the maintenance of the Second Army in 1918—details of the Second Army L. of C.—points suitable for attack on the railways—need to block supply in front of reserve depots—but far enough back to put intolerable strain on motor transport—General Plumer's recommendations for attack on German system of maintenance.

Need for new technique of supply—bulk shipments—base ports as objectives—need for base depots to be farther from the battle area—larger reserves may be necessary in the theatre of war—need for more and smaller depots—dangers of attack on ammunition depots—Campagne and Audruicq—petrol base depots—need to develop light engines burning heavy fuel—Frozen meat—results of dislocation and restriction of work in depots of all sorts—need for new lay-out of lines of communication.

Air action depends for effect on dislocation more than actual damage—need for balanced movement in a transportation system—cumulative effect of a stoppage.

RAILWAYS—*History* of railway bombing in Great War—Neuve Chapelle—conclusions arrived at in July 1915—Loos—policy laid down in February 1916—influence of factor of air superiority over battle-line in reducing intensity of railway bombing—the appreciation before Passchendaele—Foch's instructions of 1.iv.18—railway objectives allotted—the Nash memorandum—relations between Staff and technical advisers—the 'concentrated bombing' schemes—reasons for meagre results of railway bombing in the War.

Objectives in a railway system—junctions and main-line stations—Amiens during Somme—East Prussia—need for detailed knowledge of junctions—results at Thionville and Metz Sablon—trains on the move between stations—Anti-aircraft defence of trains—need for selecting sections where traffic is heaviest—keeping a block closed—repair trains and break-down gangs—control offices—system of intercommunication—signalling apparatus—the absolute block system—marshalling yards—engine sheds and workshops—effect on reserves of rolling stock—locomotives—the Austrian high-powered locomotives—the 'round-house' engine sheds—fuel reserves—watering arrangements.

ROADS—movement must be regarded as a whole—road transport has increased flexibility and lightened burden on railway—importance of roads as objectives in inverse ratio to density of railway system—undeveloped countries—the Khyber road—civilized countries mostly well roaded—roads difficult to break—possibility of blocking in towns—gas mains—supply not usually worth attacking forward of railheads—but exceptions, e.g. the Bar-le-Duc road into Verdun.

BRIDGES—Not usually good objectives—important ones well defended—big ones difficult to damage—small ones easy to repair—but may be small bridges of great strategic importance—the Kuleli Burgas bridge—normally regard bridge as defile and attack traffic crossing it—the Köln bridges during the German concentration in 1914—the Marne bridges in July 1918.

Conclusion—need for adjustment of organization to new standards.

PART III. THE BATTLE OF AMIENS—AUGUST 8TH–11TH, 1918

The situation when the plan was conceived—the development of the plan—preliminary arrangements and orders—the air instructions—the course of the Battle—August 8th—air operations—August 9th—air operations—August 10th and 11th—air operations—the German air forces—disappointing results after very successful break-in.

No object laid down for air action—5th Brigade instructions to pilots—the blue line and its relation to the plan as a whole—object suggested—employment of fighters to 'break the crust'.

The plan as it was—action of the low-flying fighters—employment of bombers—the G.H.Q. programme—neglect of available *information*—and of *technical advice*—use of bombers to put down barrage on Somme bridges—neglect of *reconnaissance*—employment of No. 48 Squadron—comments.

The plan as it might have been—an *appreciation*—*information* available—*enemy forces*—reserves—most probable routes to be used by enemy reserves—elaboration of the object—*roads*—suitable defiles—*railways*—five points to be attacked—the direct support of assaulting troops on to the blue line—*Plan* suggested.

The principle of concentration again—illustrated under four headings—Amiens front the decisive point on August 8th—*Concentration of the maximum number of squadrons on the decisive task*—British air forces in France in August 1918—the actual concentration for the battle—comparison with enemy air concentration on March 21st—comparison with tank concentration.

Disposition and employment of squadrons that did *not* take part in the battle—squadrons under the command of other Armies—the 5th Group on the Belgian coast—the attack on Varssenaire aerodrome—the Independent Force—none justifiable *at the time*—comments—the Home Defence force.

Possible air concentration suggested—its practicability from an administrative point of view—some preliminary moves of squadrons necessary—need not have given away surprise—arrangements for assault squadrons to learn the ground—this concentration possible on grounds of security in other sectors.

Development of the maximum effort by squadrons concentrated—comparison of other units by 205 squadron's standard—need to husband resources in periods of minor activity.

Concentration of effort on minimum number of objectives—Foch's instructions of 1.iv.18—number of objectives actually attacked on August 8th.

Centralized control of air operations—difficulty when operating with Allies—degree of co-ordination actually exercised on August 8th—control of R.A.F. in France

CONTENTS

decentralized among eight different commanders—one air commander should
have had control.

The actual moves of enemy reserve divisions on to their Second Army front
between August 8th and 11th.

PART IV. CONCLUSIONS

Influence of new inventions on war—three really revolutionary developments—
gun-powder—the machine-gun—the AIR.

The effect of air action on the maintenance of modern armies—examples of
defeat in the past due to maintenance difficulties—Colonel Wingfield's conclusions
—First conclusion.

Single lines of supply—the Turks in Gallipoli—the Palestine campaign—the
Suez canal—Russia and Japan in the Far East—Vladivostok and the Trans-
Siberian railway—possible operations in Manchukuo—Second conclusion.

Need for staffs to use larger maps—the Japanese air force and operations against
Vladivostok—air action from Cyprus against the Turkish rail communications in
Asia Minor—Third conclusion.

Air action and the strategic concentration of armies—the German and French
concentration in 1914—but vast railway resources of Franco-German frontier—
effect of road transportation in increasing flexibility of military movement—railways will not again be so congested—but different conditions in less developed
countries—the Budapest–Belgrade line and the concentration against Serbia—
Fourth and Fifth conclusions.

Dispositions in the concentration area—new technique necessary—Sixth conclusion.

Forward movement from the area of concentration—the Germans in 1914—
effect of demolitions in canalizing enemy communications—influence of armoured
forces on a line of communication—need to co-ordinate their action with that of
air forces—Seventh conclusion.

Influence of air action on the use of railways for minor strategic movements—
the rail moves in East Prussia 1914–15—all within range of Russian air bases—
Eighth conclusion.

The Air striking force in battle—the conversion of the 'break-in' into the 'break-
through'—must stop the enemy moving his reserves to the threatened point in time
—Ninth conclusion.

Correlation of the ground and air plans—Churchill's memorandum of 1917—
the Germans in the Château Thierry Salient, July 1918—the interdependent roles
of the two Services in winning a land battle—Tenth conclusion.

Summing up—General Conclusion—traditional armies obsolete—the Army of
a Dream.

LIST OF APPENDIXES

CONTENTS

SKETCH-MAPS (following page 222)

INDEX (page 223)

PART I
AIR SUPERIORITY

I
THE OBJECT

'The National object in war is to overcome the opponent's will. .
. . Since the armed forces are the only instruments of offence or
defence, these forces or such of them as are capable of influencing
the decision, must be overcome. The aim of the Army is therefore—in co-operation with the Navy and Air Force—to break
down the resistance of the enemy-armed forces in furtherance of
the approved plan of campaign.'[1]

THUS the object of an army in a land campaign is to defeat the enemy's
army; that of the air force contingent in the field is to assist and co-
operate with the army in the defeat of the enemy's army, and of such
air forces as may be co-operating with it. It is necessary to emphasize
this rather obvious truth in order to clear the air of a certain amount of
misunderstanding that too often in the past has obscured the issue of
this subject. The War of 1914 to 1918 in the air was, for obvious and
natural reasons, an 'Army co-operation' war in the narrow technical
sense of the term. At the outset, and for many months afterwards, the
only important use of aircraft appeared to men's minds to be that of
reconnaissance. The potentialities of the aeroplane as a means whereby
commanders could obtain information—could see the other side of the
bill—were so obvious, and the offensive capacity of aircraft in those
days was so slight, that it was only natural that their use as a positive
striking force wherewith to influence the decision should be overshad-
owed. As the War went on and the land forces on either side increased
in size and complexity, the demands on aircraft for observation for the
ever-growing mass of heavy artillery assumed the position of greatest
importance. Air bombardment, particularly against such objectives as
the enemy rail communications, was not altogether neglected even

[1] F.S.R. ii, sect. 4. I.

B

early in the war, as will be shown later in this book. But it is broadly true to say that up to the end of the war the primary object of all air operations, with one exception, was to secure air superiority over the battle line, to enable our reconnaissance and artillery aircraft to carry on their work of close co-operation. The one exception was the group of units known as the Independent Force R.A.F.; and even in the circumstances leading up to the formation of that force it is possible to see clearly the influence of this idea, this underlying policy that the primary object, the service of first importance that air forces could perform for the land forces, was the purely ancillary service of reconnaissance and observation.

Trenchard's Squadron

It would be foolish to pretend that the work of close cooperation, of those units now known as Army Co-operation Squadrons, is not of very great importance to the Army. Indeed there will very likely be occasions when to ensure an uninterrupted flow of information, or accurate and unremitting observation for artillery, will again for a time be the most valuable service that can be performed by the air forces in the field as a whole. Such a service may well be again, as it has been before, a most important factor in the defeat of an enemy. But it is the object of this book to draw attention to the other aspect of air power in land warfare, namely the positive influence which can be exerted by an air striking force in direct attack upon objectives on the ground. It is this aspect that in the opinion of the writer is still seriously underrated in the British Service to-day, although it may have an influence on the course of a campaign out of an proportion to that which can be exerted by the purely ancillary service of reconnaissance and observation.

And this leads up to the point that *in a land campaign* the primary objectives—that is to say those against which action will lead most directly to a decision—will always be the enemy *land forces*, their communications and system of supply. This point perhaps needs some elaboration. For the purpose of this book a land campaign must be taken to mean a campaign, or a stage in a campaign, of which the primary object for the time being is the defeat of an enemy army in the field. Such a stage may occur in a war in which the ultimate aim of the enemy is the reduction of this country by air measures. It has already been suggested, for instance, that the first stage of an air war against this country may be a struggle on the ground for the possession of bases in the Low Countries, from which an effective air invasion of England can be sustained. In these conditions, until the enemy army has been confined within or driven back to its own frontiers, or to a position which gives sufficient depth for the air defence of this country, the primary object of the combined forces in the field will be the defeat of that army—or at least its expulsion to the requisite distance. And that is what is meant by the expression 'land campaign'.

Once that object is achieved, then the ultimate reduction of the enemy nation may (and very likely will) be undertaken, not by the traditional methods of land invasion, or by continued assaults upon their armies in the field, but by air measures. That is to say it will become an air campaign, and the task of the army will be simply to protect the air bases. When that happens then it is arguable that a decision may be secured by action against enemy air forces—since the complete defeat of those forces would place the enemy nation at the mercy of unrestricted air action. As a matter of fact it is more likely that even in an air campaign action against air forces will remain only subsidiary, and the decision will only be gained by direct action against the hostile vital centres. This, however, is getting on to ground beyond the scope of this book, and for the present purpose it is sufficient to emphasize that in a land campaign all action other than against *military* objectives (as opposed to objectives connected with the enemy air force) is merely secondary.

This does not mean that, even in a land campaign, action against enemy air forces and their system of supply and provision can be neglected as unimportant. On the contrary, it will be absolutely essential for two main reasons: firstly to enable our own air offensive to be directed with the minimum of interference against the objectives best calculated to contribute to the primary aim of the national forces in the field; and secondly to protect our own army and its reconnaissance aircraft from hostile air interference. It is thus a means to an end, an essential measure of security upon which all offensive action must be based. But air fighting in itself, the destruction of enemy air forces, will not give us a decision in a land campaign.

Even to-day we find the tendency to exaggerate the importance of the secondary, negative object of air action reflected in our training manuals. We read, for instance, that 'the first duty of the air force contingent is to create and maintain an air situation such as will assist the army to achieve its object, and will prevent undue interference from enemy air attack'.[1] If this means the first duty in point of time it is arguable that the creation of a favourable air situation may sometimes be a necessary precedent to the attack on the primary objective, though the two processes will more commonly be simultaneous. But it is not infrequently claimed—and the claim finds some support in the manuals—that the maintenance of a favourable situation in the air is the *principal* task of both bombers and fighters in the field. This is definitely not so. Air superiority is only a means to an end and, unless it is kept in its proper place as such, is liable to lead to waste of effort and dispersion of force. Indeed, there is so much confusion of thought on this subject of air superiority in relation to the object in war that it is worth a brief general examination.

The fact is that 'air superiority' has become something of a catchword. It is easy, and sounds convincing, to say briefly 'the job of the air force is to gain air superiority' and leave it at that. But what does it mean? The official definition of the term is 'a state of moral, physical, and material superiority which enables its possessor to conduct operations against an enemy and at the same time deprive the enemy of the ability to interfere effectively by the use of his own air forces'—in other words it means the capacity to achieve our own object in the air and to stop the enemy achieving his. To 'deny to the enemy freedom of action for his aircraft' is only the second, and the negative, half of this meaning. We can learn something from the naval analogy. The object of sea supremacy is the control of sea communications, to secure them for our own use and to deny their use to the enemy. It can equally well be said that the object of air superiority is the control of air communications *firstly for our own use* and secondly to deny their use to the enemy. And the use to which we require to put them is to 'conduct operations against an enemy'; and this, in a land campaign (which is the subject of this book), means to *break down the*

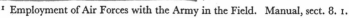

[1] Employment of Air Forces with the Army in the Field. Manual, sect. 8. 1.

resistance of the enemy army, which includes reconnaissance and observation for the army and direct attack by the air force— both directed to the same end. Another point of similarity with the naval term is interesting. In the proud old days before the War we used to talk of the Command of the Seas. To-day we have dropped that term and speak of control of sea communications—and air superiority, not command of the air. The reason is to be found in the same quality, the capacity for *evasion*, at sea of the submarine, in the air of the aeroplane. Once warfare gets into the third dimension, whether above or below the surface, the cubic area of the battle-fields is so immense that absolute command is hardly ever practicable.

All this must not be taken to mean that we can in fact control the air communications without fighting the enemy air forces. Just as at sea the surest method of achieving the object—control of sea communications—is the destruction of the enemy fighting fleet, so in the air the surest method of securing freedom of action for our own aircraft would be the destruction of the enemy's air forces, if that were possible. But as in naval warfare the most that can normally be secured under modern conditions is the neutralization of the enemy's battle fleet, so also in the air. The unlimited opportunities for evasion have the result that the most we can hope for in a war against a first-class enemy is to throw his air forces on the defensive, to neutralize them, to enable our own aircraft to work with the minimum of interference, and to reduce enemy air action against our own forces to the minimum. And even this limited result can only be secured by hard and continuous fighting.

Note, however, that in both forms of warfare, at sea and in the air, battle with the enemy's sea or air forces is only the means, not the end—or, in 'Staff duties' jargon, the Method, not the Intention. It is of far more than merely academic importance to distinguish clearly between the Method and the Intention, between the objective and the object, between the steps taken to achieve the end and the end itself. It was perhaps some confusion of thought on this head which led even Mr. Winston Churchill, in a memorandum written in October 1917 which was otherwise a masterpiece of strategical foresight,[1]

[1] App. V, Part II, *The World Crisis, 1914–18.*

to state as an axiom the following: '. . . The primary objective of our air forces becomes plainly apparent, viz. the air bases of the enemy and the consequent destruction of his air fighting forces. All other objectives, however tempting, however necessary it may be to make provision for attacking some of them, must be regarded as subordinate to this primary purpose.' Note the confusion—'objective' in the first line, 'purpose' in the last. This confusion may be in part due to the perhaps rather disproportionate amount of attention given to the operations of the fighter aircraft in the last war. For the fighters, but for the fighters alone, the destruction of enemy aircraft may be said to be the primary objective. But the fighters, though an immensely important arm, are nevertheless only one arm of the air forces; and their operations, if they could be considered apart from their effect on those of the rest of the Service, would appear as remote from reality as the jousts of medieval champions. Of course they cannot be so considered, and the influence of their activities is of vital importance—but only as a contribution to the end, and not as the end itself. The point is well explained by Major Sherman, of the U.S. Air Corps, in his able book on air warfare.[1]

'*The basic mission of pursuit aviation*:—with the exception of pursuit, all other branches of aviation have a dual role. Their missions may be of the nature of a service, as is habitual with observation aviation; or they may consist in offensive action against hostile elements on the surface of the earth, which is the proper role of bombardment and of attack aviation. In any case they have a certain duty to perform which arises from other than air considerations. In so far as the latter are concerned, their missions are almost invariably of a defensive character. Whenever consistent with the playing of the first part of their roles they studiously avoid combat. With pursuit aviation on the other hand, air combat is the sole reason for its existence. . . . The study of air warfare *in its most limited sense*,[2] is the study of pursuit strategy and tactics . . . to-day the basic mission of pursuit aviation is the destruction of all hostile aircraft and the protection of friendly aircraft. . . .

'In a certain sense, pursuit aviation may be called an auxiliary to the other branches of aviation, for the criterion of its value is the

---

[1] *Air Warfare*, by Major W. Sherman, p. 127 et seq. Note: the Americans call their Fighter Squadrons 'Pursuit Aviation'.   [2] My italics.

effect it has on the air situation; and this in turn is of importance only to the extent it affects the operation of attack, bombardment and observation aviation. . . .'

This is the point: the air situation has no importance in any form of war except in so far as it affects the situation on the ground, and the operations of those arms of the air force who are engaged against hostile objectives *upon the ground*—whether by direct attack or by indirect action in the form of reconnaissance and observation. And the fighters are an arm of the Service whose influence is entirely indirect and auxiliary,[1] although, as Major Sherman rightly goes on to point out, 'No mistake could be more destructive in its consequences than to underrate the value of pursuit aviation to air operations as a whole, under any conditions of serious warfare'.

The correct relation between air fighting—attack on the enemy's air forces—and the true object of an air force in war was well understood by Marshal Foch in the closing months of the War. His plans for the employment of the great Inter-Allied Bombing Force of the 1919 programme (of which the Independent Force R.A.F. was the nucleus) were primarily aimed at the dislocation of German war industry and munition supply; and he laid down in some detail, and in order of importance, the various targets to be bombed—such as chemical factories, industrial and commercial centres in the Rhineland and the Saar, and focal points on the railways serving those vital areas. But Sir Hugh Trenchard, who was to command the Inter-Allied Force, was given a free hand to deal as he thought fit with the bases from which enemy aircraft might operate to interfere with his work. His work, the object of his operations, was to bring pressure to bear on Germany by dislocating her centres of war industry and the rail communications serving them. In order to enable his squadrons to carry out this programme they would have to fight their way through the enemy's defensive screen of fighters, and would also have to direct some of their energies to the neutralization of enemy air forces by attack on their aerodromes—Method, not Intention: a necessary step to achieve the object, but not the object.

There thus appears at first sight to be a wide divergence between the object of the army, as stated in Field Service

[1] See p. 20 below.

Regulations, 'to break down the resistance of the enemy's armed forces', and that of the air force, as officially stated, 'to break down the enemy's resistance . . . by attacks on objectives calculated to achieve this end' (although such objectives, for the air force directly co-operating in a land campaign, happen to be the enemy land forces in the field, their communications and system of supply). The air arm seems to miss out what has always been regarded as an essential stage in war, the reduction of the enemy's corresponding armed forces. But actually the difference is more apparent than real, and is mainly a matter of degree. The soldier recognizes that in order to achieve the national object of overcoming the opponent's will it is normally necessary for him ultimately to undertake—or at least to threaten—the occupation of the enemy's country or the interruption of his vital lines of communication and supply.[1] The airman strikes direct at those objectives. And the sole reason why he is able to do so—the first important difference between air forces and armies—is that, within his tactical range, which may be as much as 300 or 400 miles, *he is independent of lines of communication and has no flanks*. To use a simplified illustration: an army setting out to invade a hostile country could not possibly afford to forge straight ahead in the direction of the enemy's capital, ignoring all hostile forces other than those which it actually meets in its path. The essence of war on land is manœuvre, and the army which allows its opponent to turn its flank and get astride its line of communication, on which it must rely for the vast mass of food, ammunition, and other stores essential to its fighting efficiency, is a defeated army. Therefore before an army can pass on to become the instrument of national policy by the occupation of the enemy's country, it is bound first to defeat the enemy's army and thus *secure its own line of communication*. An air force, on the other hand, can reach its objective without prejudice to its own security or its capacity to damage that objective when it does get there, and can return to its base through an enemy *literally astride its line of flight*. It follows, of course, that against a first-class enemy that air force will have to fight such hostile air forces as it encounters on the direct route to and from its objective; and it will often be necessary to divert some proportion of its energy to the neutralization

[1] See F.S.R. ii, sect. 4. 1.

of the enemy air forces by attacking them at their bases, in order to reduce the numbers or the morale of the hostile air forces that it *does* encounter on the way to and from its primary objectives. But the point is that an air force *can* get to those objectives, do its job, and get back again without the preliminary total defeat of the corresponding hostile armed forces—which an army cannot.

The only other important difference, in the strategical sense, between armies and air forces is closely akin to the first, and arises out of the same quality of mobility in the third dimension; and is that *an air force is not committed to any one course of action.* Once a military commander undertakes any course of action he cannot make any fundamental alteration in his plan without incurring great dislocation and delay; he is in fact committed —in an increasing degree as mechanization progresses—to the course originally selected, owing not only to the relatively slow rate of movement inherent in land forces themselves, but even more to the elaborate and cumbrous administrative machinery and system of communications upon which they must rely. Indeed it is a curious and unfortunate paradox that the petrol engine, which it is hoped will restore to modern armies the tactical mobility of which they were deprived by the machine-gun and barbed wire, at the same time must often result in seriously reducing their strategical freedom of action. A mental comparison of a modern mechanized army with Lord Roberts's columns in the second Afghan War, or with Budienny's cavalry divisions in the Soviet-Polish campaign, will confirm the truth of that.

An air force, on the other hand, is not so committed. It can switch, literally almost at a moment's notice, from one objective to another several hundred miles away, from the same base. Recent examples will be familiar to the reader of squadrons engaged in the Mohmand country one afternoon being in action from the same base against objectives in Waziristan next morning. Furthermore, a sound organization and system of aerodromes, which have the notable merit of being inexpensive, confer on an air force the added advantage of being able to change their base with the minimum of dislocation and delay— an advantage that will be still further enhanced by the development of transport aircraft.

A simple illustration will suffice to make this point clear. There was, and is still, a school of thought which held that the B.E.F. in 1914 could best be employed from a base on the Belgian coast against the flank of the German invasion. It was in the event decided to employ that force in direct co-operation with the French army, and once it had concentrated on the French left wing, it was irretrievably committed to that role. An air force under similar circumstances could have been employed at the outset from bases in Belgium against the communications of the German flank armies through Aachen and Liége; and from those same bases could still have been switched, as and when the need arose, in whole or in part, to—say—the direct assistance of General Lanrezac's army about Charleroi, or against German aerodromes in north-east Belgium if the air situation so demanded.

It must then be apparent that air superiority is not a definite condition to be achieved once and for all, a stage to be passed from which the air force can proceed to other forms of activity. It is not a phase to be gone through, a necessary preliminary to be dealt with as expeditiously as possible before the real business can begin, like the minor attacks that were sometimes made in France in order to gain ground to secure a suitable starting-line for a large-scale offensive.[1] Air superiority will be gained and will have to be constantly maintained by striking direct at those objectives which are of first importance to the enemy at the time, whatever they may be; and by persisting in this line of action against opposition and in spite of casualties, assisted in varying degree by diversions in the form of direct attack on the enemy's air forces. The struggle for air superiority is part and parcel of all air operations against a first-class enemy; and though much can be done by superior organization and equipment to provide for the physical and material factors before we go to war, the essential third factor—perhaps the most important of all—the moral factor, can only be secured by an instant and unremitting *offensive* directed against the primary objective, whatever it happens to be at the time. 'Air superiority is only a means to an end.' But it happens that to go straight for the end is the best, in fact the only sure, way of achieving the means.

[1] For the one exception to this rule see p. 39.

II

THE MAIN OFFENSIVE

'Air superiority is obtained by the combined action of bomber and fighter aircraft. The detailed measures to obtain and maintain the requisite air situation must vary with the circumstances of the campaign, but purely defensive measures will rarely be successful.'[1]

In this and the next chapter it is proposed to examine briefly those detailed measures by which we must obtain and maintain the control of air communications for our own use and deny them to that of the enemy. And in order to afford a background for this examination it may be useful to remind the reader of two periods during the Great War, in different theatres and under widely different conditions, which clearly illustrate the influence that a favourable air situation may exert upon the operations of an army. The first example—that of Palestine—provides an exception to the rule that absolute command of the air in a theatre of war is an unattainable ideal. In 1917, before General Allenby's arrival, the German air forces operating in support of the Turkish army had enjoyed a high degree of superiority in the air. The British aircraft on that front were few and of inferior performance, and the consequent unfavourable air situation was a contributory factor in our earlier reverses at Gaza, and had a generally adverse effect on the morale of the troops, none the less potent for being indefinable. One of General Allenby's first acts on assuming command in Palestine was to demand, and obtain, three additional squadrons of up-to-date aircraft. The air situation naturally took an immediate change for the better, a change which was promptly reflected in the victory of Beersheba. Allenby was a master of strategical surprise, and the success at Beersheba—like the greater victory a year later—was largely due to the adoption of various devices to lead the Turks to imagine that the attack was coming else-where, on this occasion at Gaza. Says Colonel Wavell in his book on the Palestine Campaign:[2] 'All these devices to mislead the enemy would have been of much less avail had not the new

---

[1] Manual, sect. 9. 2.   [2] *Campaigns in Palestine*, p. 107.

squadrons and more modern machines received from home enabled our air force in the late autumn to wrest from the enemy the command of the air which he had so long enjoyed in this theatre.'

But if the effect of air superiority was striking in the autumn of 1917, it was far more so a year later, when General Allenby staged his great break-through (incidentally the only break-through which a British army ever achieved in the war) that was to end the campaign in the Middle East. By this time our air superiority in Palestine practically amounted to complete command. Once more the preparations for the attack included the most elaborate measures to deceive the enemy: empty camps, rows of dummy horse-lines, and artificially raised clouds of dust at the Jordan valley end distracted the Turks' attention from the stealthy concentration of the British and Australian mounted and dismounted divisions at the Mediterranean end of the line. To quote Wavell again:[1] 'It was above all the dominance secured by our air force that enabled the concentration to be concealed. So complete was the mastery it had obtained in the air by hard fighting that by September a hostile aeroplane rarely crossed our lines at all.' During the final preparations and on the morning of the attack we had fighter patrols sitting over the enemy aerodromes, which effectively prevented any hostile aeroplane from leaving the ground[2] at all; and both close co-operation pilots and the bomber and fighter patrols that co-operated with the mounted troops in the pursuit, with such disastrous results to the enemy, were able to go about their tasks completely unhampered by hostile air opposition. Superiority on this scale, amounting as it did to absolute command, can rarely if ever be secured in operations against a first-class enemy. But we may note in passing that on this occasion not only did it endow the commander with the invaluable—and otherwise almost unattainable—advantage of strategical surprise, but also enabled an air striking force to make perhaps the most decisive contribution it has ever made to the issue of a battle by direct action against an enemy army.[3] Its effect was summed up by General Allenby in his dispatch

[1] *Campaigns in Palestine*, p. 201.
[2] *Vide* Evidence of German M.O. at Jenin, *History of the Desert Mounted Corps*.
[3] See p. 102 below.

dated June 28th, 1919, in the following words: 'The superiority established by the Air Force over the enemy was one of the great factors in the success of my troops.'

The Palestine campaign is perhaps not altogether a fair example, because in that theatre after the autumn of 1917 we were opposed in the air by an enemy numerically very inferior, whose difficulties of supply and technical maintenance, already difficult enough owing to the distance of the front from Germany, were enhanced by the nefarious activities of the British prisoner-of-war working parties at the Taurus tunnels. But on the Somme in the summer of 1916 we were up against an enemy of the first quality, yet that battle provides an example of a local superiority approximating more closely to absolute command than anything we saw in France before or since.

'The beginning and the first weeks of the Somme battle', says von Below, commanding the First German Army, 'were marked by a complete inferiority of our own air forces. The enemy's aeroplanes enjoyed complete freedom in carrying out distant reconnaissances. With the aid of aeroplane observation the hostile artillery neutralized our guns and were able to range with the most extreme accuracy on the trenches occupied by our infantry; the required data for this was provided by undisturbed trench reconnaissance and photography. . . . On the other hand our own aeroplanes only succeeded in quite exceptional cases in breaking through the hostile patrol barrage and carrying out distant reconnaissances. Our artillery machines were driven off whenever they attempted to carry out registration for our own batteries. Photographic reconnaissance could not fulfil the demands made upon it.'[1]

These words of the enemy Army Commander principally concerned describe very graphically the extent to which control of the air for the use of our own close co-operation aircraft, while denying it to those of the enemy, was secured in this great battle. Even in September, when Hindenburg had succeeded von Falkenhayn as C.G.S. and had concentrated more than one-third of the German air force on the Somme front, Sir Hugh Trenchard was still able to report to the Commander-in-Chief that 'A.A. guns have only reported 14 (German) machines as having crossed the line in the 4th Army area in the last week,

[1] Report by General von Below on the First German Army in the Somme battle, quoted in the Official History, *The War in the Air*, vol. ii, p. 270.

whereas something like two to three thousand of our machines crossed the line during the week'.

So the primary object of the air operations, to secure freedom of action for our close co-operation aircraft over the actual battle area, was attained. But this is not the whole picture. Far over the German side, out of sight of our forward infantry, very intensive air fighting was taking place throughout the whole period of the battle. The Official History describes many examples, particularly of the fighting over our bombers' objectives on the enemy lines of communication leading to the battlefield, where the enemy Fokker and Roland fighters were especially active. There was thus no command of the air corresponding to the situation in Palestine just described. We were, however, able to secure a good working local control of the air *in the area where it was of most importance at the time,* that is where our army co-operation squadrons were engaged in their task of artillery observation and close reconnaissance. And it was our great artillery superiority, itself largely rendered possible by that control, that enabled the New British armies to break in to the most formidable system of field defences that has ever faced an army in war.

*   *   *   *   *   *

'It must now be told how it was possible for the Royal Flying Corps to do its work for the Army little hindered by the German air service. This was brought about in two ways—by seeking out and fighting the enemy's aeroplanes far over his own lines, and by creating such a threat to the vitals of his communications, by incessant bombing, that he was compelled to use up much of his fighting strength in defence.'[1]

The reader who aspires to a detailed knowledge of the air operations during the Somme battle must turn to the excellent account set forth in the Official History. But the opening sentences of that account—quoted above—will serve as a text to a general examination of the methods by which air superiority must be attained and maintained. Indeed, they find an echo in the section of the Manual dealing with air superiority:

'where conditions are favourable a temporary advantage in the struggle for air superiority may be obtained by attacking enemy aerodromes. . . . Normally, air superiority is more likely to be

[1] *The War in the Air,* vol. ii, p. 251.

gained by the attack of other objectives with a view to upsetting the plans of the opposing commander, and causing him to divert aircraft for their defence.'[1]

From an analysis of the methods adopted to secure air superiority there emerge two main principles, which for the sake of clarity it may be as well to state early in this chapter. These principles are not merely academic theories of how the situation ought to work out; they are based on actual practical experience of how things really did work out in air warfare on a large scale between first-rate opponents, the British and French against the Germans, in the Great War. They are as follows:

A. Even assuming approximately equal resources on both sides in personnel, training, and equipment, air superiority can be gained and must be maintained firstly by the adoption of a resolute bombing offensive against the vital centres of the enemy. By this means he will be thrown on the defensive in a dual sense:

1. He will be forced to use up his strength in defence, to divert aircraft from their primary *task* which alone can be decisive, so that instead of striking at *our* vital centres, and thus exerting a direct influence in the decision, they will have a passive role thrust upon them.

2. By judicious selection of objectives, he may be forced to divert even these defensive aircraft from whatever may be the really decisive *area* at the time in order to protect those vital centres which—even though not actually in the area of the decisive battle—he cannot for military or political reasons afford to leave unguarded.

This course holds out the traditional advantages of the offensive, in that we grasp and retain the initiative, force the enemy aircraft to meet and fight us under conditions of our own choosing, and thus deprive them of the capacity for evasion.

B. This leads on to the second principle, which is that the offensive against the enemy's vital centres must be supplemented in varying degree by direct action against the enemy air forces. Such action will take two forms:

1. By fighters, seeking out and destroying the enemy air forces in the air, in those areas where they are most likely to be encountered. These areas, owing to the increasing

[1] Manual, sect. 9. 4.

capacity for evasion in the air, will normally be at the enemy's points of departure or of destination. That is to say, the most profitable areas will usually be over the enemy's aerodromes or over the objectives of our bombers; but they may at times be those in which our close co-operation aircraft are operating, or even in certain conditions over the objectives of enemy bombers within our own lines.

2. By bombers, supplemented sometimes by low-flying fighters, attacking the enemy air forces upon the ground, their aerodromes, bases, aircraft depots, and technical establishments.[1]

These broad principles will be considered in some detail in the following pages. But before going farther it seems necessary to make a short digression and attempt a definition. In all the literature of air warfare, official and otherwise, no words are more overworked, or often loosely used, than this convenient expression 'vital centre'. Strictly speaking a vital centre is an organ or centre in a man, an army, or a nation, the destruction or even interruption of which will be fatal to continued vitality. Note that actual *material destruction* of a vital centre is *not* essential in order to be fatal. Thus a man's windpipe is a vital centre; yet it is not necessary to cut it but only temporarily to stop air getting through it in order to kill that man. One or more essential railway junctions may be vital centres of an army in the field; yet it is not necessary absolutely to demolish those junctions, but only to prevent railway trains passing through them for a sufficient length of time, to be fatal to that army. The same applies to the blast furnace producing steel for ships or guns, the docks where military personnel or stores are handled, or the factory producing magnetos for aeroplanes. The point is that a vital centre is some organization of which a stoppage or even a sufficient restriction of output or operation will be fatal. So a convenient definition of the term 'vital centre', within the narrow limits of an actual theatre of war, is 'Any point in the enemy's system of supply or communication of which the destruction or interruption for a sufficient length of time will, either immediately or in due course, be fatal to his continuance

---

[1] In this chapter we are considering only action against enemy aerodromes and technical air establishments in the field. Bombardment of aircraft factories will, of course, have an important effect on air superiority in the long run.

of effective operations'. But unfortunately we cannot leave it at this—we cannot confine ourselves strictly to the literal interpretation. War is a human activity, and, human nature being what it is, decisions in war are influenced by factors other than those of cold military expediency. The statesman has to consider, and so the soldier must take into account, many factors, moral, political, social, economic—he has in fact to take the national point of view, not merely the military. So that some organization, some centre there will be of which even the total destruction would not be fatal to the continuance of military operations if nations were composed of robots; but which nevertheless is humanly so important that no government can afford to leave it unprotected; which in fact as far as air operations are concerned becomes a vital centre, in that it has to be treated in the same way as a literally vital centre and has equally to be protected against interference.

It is difficult to be more definite about a subject which is necessarily so imponderable. It must be sufficient to quote by way of illustration the example of London in the last war. Despite the importance of Woolwich arsenal and other centres of munitions industry in the London area, there is no doubt that we should not have had to tie up such large numbers of aircraft in the defence of London against the very weak and sporadic attacks directed on it, but for its importance as the capital city and greatest centre of population, commerce, and finance. The bombing of London at that time could not possibly have led to the defeat of our forces in the field, nor of our fleets at sea, but the Government simply could not afford to leave London unprotected or inadequately guarded, in view of the possible social and political results which might follow a serious attack, however unlikely it may have been. So when we use the term 'vital centre' it must cover not only the literally vital centres of communications in the field, or of munition-production in the industrial areas at home, but also those centres which for political or social reasons have to be treated in the same way, and afforded similar protection.

To return to the first principle enunciated on p. 15. 'Air superiority can be gained and must be maintained by the adoption of a resolute bombing offensive against the vital centres of the enemy.' The moral advantages of an offensive policy in any form of war will scarcely be questioned. And whereas modern small arms and field defences have, at least temporarily, lent superiority to the defence upon the ground, in the air all modern developments—such as cloud flying and very high speeds—are tending more and more to weigh the scales in favour of the attack. This being so, it may be argued that both belligerents will persist ruthlessly in their air offensives, each resolutely refusing to be diverted to a defensive policy and avoiding battle with the opposing air forces, until a sort of deadlock ensues. The problem is one on which it is dangerous to dogmatize. A mere bald-headed unreasoning offensive, simply for the sake of taking the offensive, is unlikely to be any more effective to-day in the air than it was in 1914 on the ground. We have but little practical experience on which to base our conclusions. It is more than likely that each belligerent will begin by an air offensive against the vital centres of the other; and it seems not improbable that the situation may go through a stage of deadlock. It does not, however, appear likely that this will be more than a temporary phase; a variety of factors, physical, psychological, and material, will come into play. National characteristics, the morale and endurance of pilots, the efficiency of aircraft and armaments, the capacity for the replacement of wastage in personnel and aircraft, the qualities of command and direction, all will tend sooner or later to tip the scales in favour of one or the other belligerent. Apart from these factors there is another which may be of even greater importance, especially in a land campaign: A very serious handicap will be imposed upon that side whose essential communications and machinery of administration and supply are most concentrated —which has in fact the *fewer vital centres*, and which is therefore correspondingly more vulnerable to attack on those centres. The extreme example of this is, of course, the army which is dependent upon a single line of supply.

It is difficult to be more definite on this subject. Our experience is limited to the conditions of a past war which will inevitably be very different from those of a future one. It must be sufficient to say that that experience, such as it is, does provide good grounds for the belief that sooner or later the side whose offensive is most intelligently directed, which is superior in the art of command, and whose morale, discipline, technique, and material efficiency is the higher—not only in the armed forces but in the nation as a whole—will surely, though possibly slowly, begin to impose its will upon the other. The civilian authorities in the towns, munition-factories, and industrial areas at home, and the Transportation and Supply Services on the Line of Communications and at the Base, may quite naturally fail to appreciate that the best way of affording them protection is to persist in some remote offensive; or to understand that, however great a degree of superiority we may enjoy in the air, it is impossible completely to deny the air to a determined enemy. Experience on both sides in the last war clearly showed that it is always difficult and often impossible for a Government or a commander in the field to resist demands from these sources for close protection. The German High Command— no amateurs in the art of air warfare even in those early days— were 'firmly convinced of the great importance of methodical bombing'.[1] Yet at Verdun, and again during the Somme battle, they were compelled by the demands for close protection against the Allied bombing offensive to use their bombing squadrons— and sometimes even reconnaissance units—defensively, on what they called *Sperrefliegen*, or barrage patrols. So the aircraft that might have been hitting us, bombing our vital centres, and intervening effectively in the operations on the ground, were thrown back upon an inevitably ineffective defensive. 'The barrage has precedence over all other work'—so ran the German instructions at the time.

'These squadrons (the Kampfgeschwader or Bomber-fighter squadrons) could not be effectively employed at the beginning of the battle of the Somme owing to the extremely difficult tactical situation in the air. . . . A rigid defensive, by means of patrols flying parallel to the front, had proved ineffective; the patrol aeroplanes being unable to prevent the enemy squadrons from crossing the line. These squadrons were thus incapable of having much influence on the aerial war.

'Further the patrol and barrage duties of the Kampf squadrons prevented them from carrying out their proper duty of bombing. The importance of successful bombing was thoroughly recognized, but the conditions on the Somme made it necessary to desist from

[1] Extract from memorandum supplied by the Reichsarchiv, Berlin. See Appendix VII to *The War in the Air*, vol. ii.

such attacks, owing to the more pressing need for the protection of artillery aircraft.'[1]

In short 'This notion . . . had the most disastrous influence on the methods of use of the airmen', as von Hoeppner has admitted;[2] and the whole story is convincing proof not only of the inherent superiority of the offensive as a policy in the air, but also of the fact that if persisted in and directed upon sound lines it will in the end compel the enemy to resort to ineffectual and uneconomic methods of protection.

It has been argued against this policy that the bomber is so essentially offensive in character that a commander in fact could not use it for defensive purposes even were he so inclined. But this is overstating the case. To begin with, even if we can compel the enemy to divert his bombing activities against *our own aerodromes* we have gone some way to accomplish our end, since the result will be that attack on our more vital centres will be correspondingly reduced. But the history of 1916 has shown that in fact bombers and reconnaissance aircraft have in the past been used actually as fighters. And it is important to remember that the modern high-performance day-bomber or reconnaissance aeroplane makes a very fine fighter; in the last war the most formidable 'aircraft destroyers' on either side were the two-seater fighters, the Bristol and the Hannoveraner, as statistics since the War have shown. In the British Service to-day one basic type of aeroplane is used in day-bomber and army co-operation squadrons, and there are even two squadrons of this same type—the Demon—among the fighters of the London defences. So that there is no doubt that if a commander in the future is compelled to yield to pressure for close fighter protection, he will find the suitable material ready to his hand in the bomber and reconnaissance squadrons.

But there is another, more 'long-term', result of a persistently offensive policy in the air. The only aeroplanes which exert a direct influence on the issue of a land campaign are those which are employed on what may be called *direct action*, either on observation for the army—to enable the other arms to strike —or striking themselves, with the bomb or the machine-gun, against objectives on the ground.[3] Therefore the more aircraft

---

[1] Appendix VII to *The War in the Air*, vol. ii.   [2] See *The War in the Air*, vol. ii, p. 167.
[3] An artillery analogy is the contrast between the field gun—which is a *direct*

we can divert to the indirect role—that is to the protection of the direct action aircraft—the more successfully are we achieving the second part of the definition of air superiority 'to deprive the enemy of the ability to intervene *effectively* by the use of his own air force'. We reduce in effect the strength of the enemy air force in so far as it can influence the operations on the ground. The first manifestation of this principle was, of course, in the beginning of formation flying in war. Before the Fokker period of the autumn of 1915 and winter of 1915–16 nearly all air work on either side had been done by single aircraft. But the result of the technical superiority and offensive tactics of the Fokkers was an order by the R.F.C. Command in January 1916 which 'laid down as a hard and fast rule that a machine proceeding on reconnaissance must be escorted by at least three other fighting machines'—here at once you have a proportion of three indirect to one direct action aeroplane. The Official Historian[1] clearly shows how the result of this change in tactics was equivalent to a shrinkage of effective strength. In the period of German predominance—the Albatross D.V. or Richthofen period—in the spring of 1917,

> 'Fifteen aeroplanes in support of three operating cameras was common. Bombing had to be curtailed. Twelve Fighters—six two-seaters as close escort and six single seaters in formation above—were sent out with a bombing formation of six B.E. aeroplanes. . . . Pilots were too much in demand for more urgent duties to allow of bombing on a scale comparable with that undertaken as an integral part of the air offensive waged during the Somme battles.'

As the war progressed this condition increased in scope and exerted an influence on the programme of building and expansion of both belligerents. The aircraft factories of both sides had an increasing proportion of their energies diverted to the production of fighters instead of direct action types of aeroplanes. For instance, by August 1918, out of a total of 2,385 German aircraft on the Western front 1,326, or about 55 per cent., were fighters, as compared with 771 out of 1,757, or about 44 per cent., on the British side.[2] The fact that the 400-odd

action weapon in that it actually hits the enemy on the ground—and the A.A. gun which is a merely protective weapon and exerts no direct influence on the land battle.   [1] *The War in the Air*, vol. ii, p. 157.
[2] The German figures are for aircraft on establishment, the British for aircraft

aircraft, mostly fighters, locked up in the defence of Great Britain, brought the British proportion of 'indirect action' aircraft up to about the same as the German was of course due to the fact that the vitals of Britain were open to German attack,[1] whereas the great majority of German vital centres were out of range of British bombers. And it can fairly be claimed that the higher proportion of 'indirect action' aircraft maintained by the enemy on the Western front is an index of the value of our consistently offensive policy.

This leads us logically to a consideration of the second aspect of the principle stated on p. 15.

It may, perhaps, have occurred to the reader that this first result of the air offensive—the imposition of a defensive role upon the enemy air force—though effective as a measure of security by protecting our own vital centres or reconnaissance aircraft from serious interference, makes no positive contribution to the interruption of the enemy's vital services, nor increases the efficacy of the means adopted to that end: That although a prolonged defensive attitude must in the long run have seriously adverse effects on the morale of the defending airmen, yet for some time at least a large concentration of defending fighters over the objectives must occasion heavy losses to the attackers, and seriously prejudice the accuracy and efficacy of their bombing. It is not to be denied that a highly organized defensive system will inflict some—and often serious —losses upon the attackers. But the criticism in general does not take into account the fact explained on a previous page that no nation at war—with possibly in some circumstances the unfortunate exception of ourselves—has any one single centre of which the paralysis by an enemy would be fatal. If there were any one single vital centre it might be possible so highly to organize its defences as at least to make the attack so costly that the attacker would not continue to face the losses involved. London is a case in point, and it might be necessary in another European war to concentrate all our resources on the defence

actually serviceable, but this does not affect the proportion. The figures for fighters on both sides include fighters employed on low-flying attack.

[1] Note that the possibility of attack on this country involves the maintenance in peace of a very large force of 'indirect action' aircraft in the Fighter Squadrons of A.D.G.B.

of London and to undertake no other military commitment unless and until its security is assured. But, with this one exception, there are degrees of importance even among 'vital' centres, degrees varying from time to time according to the strategic situation. It is a proper appreciation of this fact that will enable a commander to exploit to the full the unique capacity of air power for effective *diversion*—diversion, that is, which will contain enemy detachments out of all proportion to the effort involved.

It is, perhaps, hardly necessary to elaborate the inherent advantage conferred by extreme tactical mobility upon an attacker in the air. The attacker is in effect always on interior lines. A simplified illustration will suffice: Suppose a bomber force, with a radius of action of 300 miles, to be stationed at such a point that it can cover almost a full semicircle of enemy territory—as for instance a German force based on Roye in the summer of 1918 could have done. That force could reach any point on a circumference of nearly a thousand miles, or in an area of about 70,000 square miles; and elements of it could attack two objectives 600 miles apart simultaneously. It requires no expert geometrician to realize that even if defending fighters had equivalent tactical range, even if they could get information the moment the bombers left the ground, and even if a given number of fighters were a match for the same number of bombers (which is at least open to argument), it would still be necessary to have at least two defending fighters somewhere within the semicircle for every one potential attacker. This is admittedly an extreme case, but the postulated conditions do not, of course, prevail in reality. In actual fact when important issues have been at stake it has been found necessary in the past to have available for the defence far more than double the numbers of potential attackers; nor has anything happened to lead us to suppose that the proportion will be reduced in future —rather the reverse. The classic instance of this in war is that of the German aeroplane attacks on London; these attacks were not originally intended as a diversion, although the German commanders were quick to realize their value as such. They were, indeed, one of the most successful diversions in history, if the measure of success is the size of the detachment involved in relation to the enemy force it contains. At no time did the

German units employed against England exceed 43 aircraft—
to which must be added the appropriate small number of training aircraft. That force, based on Belgium, contained in England for home defence about 600 British aircraft, of which
nearly 400 were first line, the remainder being in training units.
It is true that there was, in addition, the possibility of a revival
of the airship raids, but to set against that there is the fact that
by no means the whole effort of the German bombing force was
directed against objectives in England. As a conservative estimate it is probably fair to say that the proportion of military
effort expended by attacker and defender respectively was as
one is to eight—a successful detachment if ever there was one.
The presence on the decisive front in France of these 400 fighters
locked up in home defence could hardly have failed to enable us
to sweep the sky clear of enemy aircraft.

Nor is this the whole story. So urgent was the political
demand for more adequate protection at home, that in the summer of 1917 at a most critical period, just before the opening of
the third battle of Ypres, Sir Douglas Haig actually had to send
home from the front in France two of his best fighter squadrons
for the defence of London. Supplies of Sopwith Camels—then
the newest type of fighter—which were intended for the re-
equipment of squadrons in France, where they were badly
needed to cope with the new German Albatross fighters, were
diverted instead to re-equip home defence squadrons; while in
France we were compelled to divert the energies of many bombers
from objectives calculated to have a direct influence on the important operations then impending, to attack the aerodromes of
the enemy night-bombers at St. Denis Westrem and Gontrode.[1]

So the activities of these few German bombers, miles away
from the scene of the battle on the ground, had an effect upon
the air situation over the decisive front which, though in-
calculable, must have been enormous; at least they prevented
us from obtaining a degree of air superiority that in all probability would have materially shortened the War. This, perhaps,
was an extreme instance, and it does not necessarily follow that
the same ratio of defensive to offensive effort will always prevail
in the air. The number of fighter aircraft required to afford
reasonable protection to a given area is governed mainly by

[1] See *The War in the Air*, vol. v, p. 58 and pp. 152–9.

the size and configuration of that area, and only in a lesser degree by the potential scale of attack. No doubt partly for this reason our own bombers of the Independent Force in 1918, though again they constituted at times a very effective diversion, only drew off from other fronts about five times their own number of German aircraft for the defence of the Rhineland industrial towns. But we as a nation are perhaps unfortunately placed in this respect. This factor of home defence will always be with us in a war on the Continent. Whether the tactical defensive gains or loses in efficacy relative to the power of offence in the air—and at the moment the latter seems more probable—it is certain that we shall never be able to dispatch fighters to assist in the defence of a European ally so long as our own country remains open to attack. As Sir William Robertson has said, 'the requirements of Home Defence, whether on land, at sea, or in the air, will, except perhaps in the case of a great crisis such as that which occurred in March 1918, invariably have to be given precedence over requirements connected with operations abroad'.[1]

The moral is, of course, that an air force commander must deliberately make use of this extraordinary capacity for effective diversion to attain air superiority *at the really decisive point*. He must exploit the extreme flexibility, the high tactical mobility, and the supreme offensive quality inherent in air forces, to mystify and mislead his enemy, and so to threaten his various vital centres as to compel him to be dangerously weak at the point which is *really decisive* at the time. There is nothing new about this: it was done in the War, and the Official History contains several examples—particularly during the battle of the Somme—of operations directed against objectives in Belgium, designed to prevent the enemy from reinforcing his air units on the Somme at the expense of other army fronts. But these incidents, interesting and important though they were, serve only as a bare indication of what now is possible in this direction. The selection of the correct objectives for this diversionary action, those centres which though vital are yet temporarily lower in the scale of decisive influence, presupposes the most exhaustive resources in information and the most meticulous system of day-to-day intelligence—not only military intelligence in the

[1] *Soldiers and Statesmen*, vol. ii, pp. 1-18.

narrower sense but also political and industrial intelligence. The conduct of the offensive calls for the highest qualities of command and the exercise of the nicest judgement, to gauge the minimum number of objectives calculated to achieve the end, to strike the proper compromise between dispersion and concentration, to assess the proper strength of the detachments necessary to exert the requisite pressure on the objectives selected, the degree of importance likely to be attached by the enemy to their defence, and consequently the strength of the forces likely to be contained. And finally the results achieved will be enhanced by real and intelligent co-operation on the part not only of the other Fighting Services but also of other Departments of State, such as the Foreign Office, the Board of Trade, and the Ministry of Propaganda or its equivalent.

In passing, the writer cannot refrain from giving an airing to a favourite bee which buzzes most persistently in his bonnet; namely the unwisdom of maintaining three separate and distinct intelligence branches in the three Service ministries—each too frequently arriving, largely from the same sources, at different conclusions to suit its own particular theory. This question of the selection of objectives is only one of many examples of the crying need for the combination of the three Intelligence branches, together with those of the Foreign Office and the Board of Trade.

So much for the first principle in the conduct of the struggle for air superiority. But before proceeding to an examination of the second, which is that this *offensive against the vital centres of the enemy must be supplemented in varying degree by direct action against hostile air forces*, it is desirable again to emphasize that such action, though vitally important, is yet from the broad strategical viewpoint a diversion. It may be as well to anticipate possible criticism on this head by admitting at once that it is a form of diversion which may—according to the strength, efficiency, and fighting value of the enemy—at times absorb the greater part, if not the whole, of our own air effort. Nevertheless it remains a diversion. It is also true that it is a diversion to which we may frequently be compelled to resort. It has been explained in the previous chapter[1] that air superiority is not a definite tangible condition which, once attained, can be left to

[1] See p. 10 above.

look after itself. In a war of any duration the air situation may be subject to a succession of remarkable—even violent—variations, such as those which marked the war in the air on the western front. All sorts of factors will tend to produce such variations. One side or the other may produce a surprise in the sphere of material—a very superior type of aeroplane, for instance. The effect of this factor has already been noted in the result on the air situation of the arrival in Palestine of the new squadrons of up-to-date aircraft. And it was to a great extent the appearance on the German side of a very superior type of fighter in the spring of 1917 that occasioned, perhaps, the most notable reversal of British fortunes in the air; to the extent that within nine months of our attack on the Somme we had almost completely lost the air superiority we enjoyed in that battle. Another important cause of the German recovery in 1917 was a change of command, with a consequent improvement in the training and higher direction of the German air force.[1] Or again, that very important, though imponderable and mercurial quality, individual morale, may be elevated or depressed by many influences such as a run of bad luck, or the almost legendary prestige and example of a few outstanding individuals like Mannock, Ball, or Guynemeyer on the Allied side, Richthofen or Boelcke on the German. In fact, between opponents of approximately equal efficiency and valour, a definite superiority in the air may never be finally consolidated until one side or the other is able by some means to acquire a great superiority in numbers, such as that which the Allies enjoyed on the western front in the closing stages of the War.

Apart from factors such as these, there is another of a different character by which we may be compelled to divert temporarily even the whole of our air effort to action against hostile air forces. There may be periods in a campaign when our own army is temporarily so vulnerable to air action relatively to that of the enemy, that he may be able to afford to accept inter-

[1] General von Hoeppner was a very able and experienced officer, a member of the Great General Staff, who had been Chief of Staff of an army in the invasion of 1914. His appointment to command the German air force after their defeats on the Somme resulted in an extraordinary change for the better in their training and higher direction. Indeed, his strategical handling of the enemy air force—considering the very limited experience of air warfare then available—could hardly have been better.

ference with his own communications, as being completely out-weighed by the importance of the fatal damage he can inflict upon ours. Such an occasion might be, for instance, while our army is passing through some dangerous defile; or while we are operating at the end of a vital single line of communications within range of enemy bombers; or during the final phase of an opposed landing, when the troops are closely packed in crowded transports or in open tows. In such conditions the factor of security assumes the first importance, and it may well be that we shall have temporarily to divert our whole air strength to deal with the enemy air force. Before undertaking a campaign in which conditions of this nature are more than a temporary phase, it will be necessary to face the fact that one arm of our striking force, the air arm, will be permanently compromised. And it should be the aim of our war policy to endeavour, if humanly possible, to avoid getting involved in a commitment in which our combined strategy is subject to such a damaging restriction.

Enough has been said here and in a previous chapter to emphasize the fact that action against enemy air forces in a land campaign is a diversion and a measure of security—and never the object. It has already been suggested that a stage may be reached sooner or later in a war when the defeat of the enemy's army in the field is no longer the primary object, or when his air forces form so large and important a proportion of his total fighting strength that the attainment of a real superiority in the air will, in itself, be sufficient to induce him to accept our terms. But these are not the conditions which it is the object of this book to examine; and in a land campaign as defined on p. 2 it must be obvious that—to put it crudely—one can ulti-mately only attain the object by hitting the enemy on the ground. And until our air forces can attain such mastery over the enemy in the air as will enable them to direct their attention to hitting the enemy on the ground, they are making no *direct* contribu-tion to the attainment of the object of the national forces in the field.

The really important thing to understand—and the point which the writer is so presumptuous as to believe is rather widely misunderstood—is the method of approach to the problem, the proper relation of action against air forces to the air plan as a whole. A concrete illustration will best explain the point, and for this purpose it will be convenient to select the hypothetical example quoted in a later chapter[1] of the air plan which might have been made for the battle of August 8th, 1918.

First it must be assumed that the air force commander, from his knowledge of the air situation generally and his experience of operations up to date, is in a position to form a reasonable estimate of the force he is likely to be able to make available for action against objectives on the ground, and consequently of the approximate scope of the task he will be justified in undertaking. Based on this estimate let it be assumed that the object laid down for him by the Commander-in-Chief in this case is briefly 'to isolate the area Bapaume–Le Catelet–Guise–La Fère–Noyon from enemy reinforcement and supply, in order to enable the attack of the Fourth Army to be carried to a complete decision'. The subsequent stages in the evolution of the plan to give effect to that object would be somewhat on the following lines.

Two parallel examinations of the problem would have to be undertaken by the Air and General Staffs in conjunction; they would be going on simultaneously, they are to a great extent interdependent, and in practice cannot be separated into watertight compartments. But for the sake of clarity they are here set down as two distinct proceedings.

An appreciation has to be made of the situation on the ground as it affects or is affected by the air plan; the detailed objectives are selected and an estimate made of the force required to produce the desired results. This part of the problem is dealt with in detail in a later chapter, and will not be further elaborated here. Suffice it to say for the sake of simplicity that it is decided to cut the railways at five points and to attack enemy columns on the march in the area described on p. 178.

At the same time the air force commander has to ask himself the following question: 'How can I create the necessary air situation to enable me (*a*) to make certain of cutting and keeping cut the railways at those points, and effectively breaking up any columns I find on the roads, and (*b*) to ensure the security of the Fourth Army and its close co-operation aeroplanes against serious interference by the enemy air forces?' This

[1] See p. 179 below.

general problem resolves itself into three more or less distinct subsidiary questions.

(a) What objectives are there at a distance from the decisive area, say in Belgium or the Rhineland, attack on which will force the enemy to employ aircraft in their defence? And are those objectives so important to the enemy, that the forces likely to be contained in their defence exceed the necessary detachments from my own force by a margin which will make such detachments worth while?

(b) In view of the relative strength and efficiency of my own forces and those of the enemy, to what extent can I depend upon the operations of my striking force against the enemy's communications to impose a defensive attitude upon the enemy air forces, and thus in themselves ensure the security of the Fourth Army against undue interference?

(c) (dependent partly upon the answers to (a) and (b)) What proportion of my force must I divert, and for how long, to direct action against enemy air forces in order (i) to make reasonably certain that my striking force will be sufficiently free from enemy air interference to enable them to perform their tasks effectively and with the minimum of loss, and (ii) to afford the necessary margin of security to the army over and above that which will be afforded by the factor in (b) above?

These are formidable problems, requiring the exercise of the nicest judgement. It is fortunate that, to compensate for the extreme difficulty of arriving in advance at an accurate estimate on any of them, we have the advantage of the extreme tactical flexibility of air power; which means that we are not committed to any one course, nor to a fixed allotment of any proportion of our force to any particular task, but can switch our strength or any proportion of it from one objective or task to another as the need arises.

Finally, having arrived at as accurate an answer as possible to the last question, and if necessary having modified the object to conform with the more modest resources which this subsequent examination may have proved to be available for its attainment, the air force commander must then decide the methods he is going to adopt to deal with the enemy air forces, on the lines which the following pages are intended to describe.

III

THE SUPPLEMENTARY OFFENSIVE—THE DESTRUCTION OR NEUTRALIZATION OF ENEMY AIR FORCES

THE strategy of the offensive against enemy air forces is a subject on which it is difficult and more than usually unwise to be dogmatic. The last War provided us with a great deal of experience of air fighting, but the conditions in which that experience was gained were mostly the narrow specialized conditions of trench warfare. The opposing armies were nearly always in close contact, and there was one quite clearly defined and narrowly circumscribed area within which local air superiority was required, the area within a few miles each side of the trench-lines. We have practically no experience of air operations in open warfare; and in particular we have never been faced in war with the difficult problem of the employment of fighters in the vital opening stages of a campaign when the armies are still many miles distant from each other. This is not to say that none of the lessons of the last War are of value or applicable to different conditions; but merely to emphasize the need for imagination, and for the capacity to adapt our methods quickly, if war comes, to conditions for which we have no historical precedent, and which may be very different from what we have imagined them; and this capacity is dependent on a sound grasp of principles. We cannot possibly draw up definite plans, or even evolve any very hard and fast tactical method to meet every possible class of contingency; but if we have a clear idea of the basic principles we shall be able to apply and adjust them to suit each contingency as and when it arises. And for a study of these principles the last War does provide some foundation.

The problem is how to 'deprive the enemy of the ability to interfere effectively by the use of his own air forces'. The ideal method obviously would be to destroy the hostile aircraft, either in the air or on the ground. But since it will usually be impracticable to make certain of destroying the aircraft themselves to a sufficient extent, our efforts in that direction must be supplemented by action to dislocate and disorganize the aerodromes, workshops, and depots on which they must rely for maintenance, equipment, and supply. The next most effective method to the actual destruction of an aeroplane is to stop it flying for lack of spare parts, fuel, or technical maintenance. So it will be apparent that action against enemy air forces is a joint responsibility of both fighters and bombers. The first part of this task, the destruction of hostile aircraft themselves, is the primary role of the fighters, and will be considered first in this chapter. After that it will be necessary to examine the supplementary action against their ground organization, which is mainly an affair for the bombers—though neither division of the task is exclusively the province of either class of squadron, and the efforts of each class must be supplemented by those of the other, as will appear from the following pages.

Coming now to the actual destruction of enemy aircraft in combat, it is proposed to follow the procedure adopted in the previous chapter, to state certain principles and then to examine them in some detail. The writer has heard it claimed as a property peculiar to air forces that owing to their high mobility and consequent capacity for evasion they can only be brought to battle by consent. This point has even been developed so far as the claim that, like surface fleets in the past, air forces could refuse contact and remain in being, a constant threat to their opponents. This in fact is only partly true, and the attempt to claim for air forces a quality which they can only very rarely possess, and to compare them with naval forces under different —and now largely obsolete—conditions, has to some extent clouded the issue, and created some misunderstanding of the problem which it is the object of this chapter to examine.

In point of fact the *strategical* conditions of air warfare, in the limited sense of combat between air forces, do not differ in any important point of principle from those governing the action of forces on land. It could be claimed that an army can only be brought to battle by consent. An army *could* refuse battle, but by doing so indefinitely it would normally be failing to implement the object of its existence, which is to enforce the national policy. Thus a defending army may decline battle with an invader; it may retire from one position to another, it may so manoeuvre as to constitute a threat to the communications of the invader and compel him to deflect his advance from vital areas—and when this is done it may again retire. Eventually, however, it will almost always be necessary, if vital national interests are to be secured, for the defending army to stand and fight. A belligerent nation may be able to afford to accept damage to the national interests to a degree which is beyond the capacity of the invader to impose, and which still is not vital—as the Russians were able in 1812; or, when the national interests are hopelessly prejudiced, a guerrilla army may for a time continue seriously to embarrass the invader without accepting battle, as the Boers did in the later stages of the South African war. But these partial exceptions do not invalidate the general rule. A defending army which refuses trial by battle more or less by consent can be brought to action by the attacker threatening *either* some objective such as a capital city or industrial area which, for reasons of national policy, the defender cannot afford to lose, *or* the communications, railways, ports, and base depots upon which the defending army itself depends for its continued existence as a fighting organization.

There was in the past one important difference between the conditions of war on land and at sea. At sea it used to be possible for a fleet to refuse combat, and to withdraw into the secure shelter of a fortified harbour—to adopt in fact the strategy of the 'Fleet in being'. But to be effective this form of strategy presupposed the existence of three conditions, none of which ordinarily apply to an air force. Firstly, it could only be adopted by the fleet of a belligerent who had not himself to depend for any vital interest, such as the national food supply, on the continued use of sea communications. Secondly, the enemy must be compelled by the adoption of this policy to 'restrict his operations, otherwise possible, until that fleet can be destroyed or neutralized'.[1] Mahan considered that even in the past 'The probable value of a "Fleet in being" has been much over-stated; for even at the best the game of evasion, which this is, if persisted in can have but one issue. The superior force will in the end run the inferior to earth.' To-day it is probably not too much to claim that the theory of the 'Fleet in being' as an element in naval strategy is a matter of history only, since the advent of the aeroplane has cancelled the third condition upon

[1] Mahan, *Naval Warfare*, p. 243.

D

which it depended for effect, namely the capacity of the inferior fleet to find complete security in fortified harbours.

From this rather lengthy digression it appears that in fact, contrary to the generally accepted theory, air forces are less able to decline actual combat than are armies or navies. At sea or on land a Fleet in being or an army at large on a flank could make itself inaccessible to attack, and at the same time could— at least temporarily—exert some influence on the enemy merely by constituting a threat. An air force on the ground is merely a mass of inert mechanism whose opponent, having no flanks, is not 'compelled to restrict his operations, otherwise possible' until that air force can be destroyed or neutralized—although this general rule, like all good general rules, has an exception which will be dealt with later.

At the risk of wearisome repetition it must again be emphasized that in this chapter we are only considering the narrower aspect of air warfare, that of direct action against enemy air forces. It has already been explained in a previous chapter that the main bombing offensive can be launched without the preliminary defeat of the hostile air forces; and the subject now under review is only the problem of the necessary supplementary action against those air forces to enable the main offensive to be effectively pursued—the problem in fact of whether air forces can be brought to action against their will, and if so by what means. The foregoing brief outline of some well-known conditions of war at sea and on land is of value in assisting us to formulate certain simple principles of air warfare, for which experience in the air during the last War provides useful confirmation and illustration. These principles can be stated as follows:

1. Air forces can only be destroyed or neutralized effectively by the adoption of an active and persistent offensive in the air— whether the underlying strategical policy is in itself offensive or defensive. Firstly, because of the immense importance of the moral factor in air fighting, and the immediately unfavourable reaction on the fighting value of an air force resulting from a defensive attitude. And secondly, because aircraft have no definite physical stopping-power, comparable to that of fortifications and field defences on land; so great is the capacity for

1-3 : principle of the active offense

evasion in the three-dimensional battle-fields of the air, so difficult is it in many conditions of weather even to ensure contact between air forces which may be seeking it, that any system of barrage or line patrols, even though they may be over enemy territory, is always uneconomical and usually ineffective.

2. This active offensive must obviously be intelligently controlled, and attacks must be directed to those areas where enemy aircraft are most likely to be encountered. The most fruitful areas from this point of view will be (in order of importance) the enemy's areas either of departure or of destination. The first and most obvious place in which to seek out the enemy will be over his aerodromes; the positions of these will usually be known, and usually the most certain way of gaining contact with a willing enemy will be by offensive patrols over the enemy aerodromes, on the lines which were such a feature of air fighting in France.

3. On the other hand offensive patrols over his aerodromes may not in themselves be an adequate means of bringing the enemy to action. The principal reason for this form of action being to contribute as effectively as possible to the success of our main bombing offensive, we may find that enemy aerodrome patrols are too indirect and incomplete a form of support to our main striking force. Our fighters may miss the enemy over his aerodromes, or his units may be so well dispersed on the ground that we cannot bring more than a small proportion of them to action, leaving the remainder free to operate elsewhere. So aerodrome patrols may have to be supplemented by others sent to the next most likely place of contact, namely over the enemy's most probable area of *destination*. This opens up a rather wider field. An accurate appreciation of where—other than at his aerodrome—the enemy is most likely to be encountered is obviously more difficult to make, and the solution will vary with the circumstances.

4. If our main bombing offensive is being soundly directed, if it is really hitting the enemy in his vitals, a certain area of destination at least for the enemy fighters will be that area where our bombing force is operating; and the more effectively that main offensive is fulfilling its object the more aircraft are we likely to meet in that area, including ultimately other classes of aircraft being used as fighters. So we can be tolerably certain

of forcing an action by sending fighters to rendezvous with our bombers over their objectives, or, if those objectives are out of range, by escorting the bombers on their outward or meeting them on their return journeys.

5. But there may be special occasions when our bomber objectives are not the most vital areas from the enemy's point of view. He—or we—may, for instance, be initiating some special operation on the ground, and it may temporarily be of the first importance to him to prevent our reconnaissance or artillery-observation aircraft working. If this is so, we can then be certain of meeting him in strength in the area where our army co-operation aircraft are engaged, in the immediate vicinity of the battle-front on the ground. We must be especially careful here not to overstep the mark and yield to the temptation to try merely defensive patrols. The test must be 'Are we most likely to get a fight and destroy enemy aircraft in this area?'; and offensive patrols in the vicinity of the ground battle must be reserved for special occasions, and must be combined with patrols over the enemy aerodromes.

It will be observed that so far we have been considering mainly the methods of bringing to action enemy *fighters*, and such aircraft as may be compelled by the success of our bombing offensive to act as fighters. But, although by destroying enemy fighters our own are most directly assisting in our main bombing offensive, still we have also to consider the very important requirement of protection for our own army and its communications, which involves the destruction of the enemy *bombers*. As already explained, historical precedent gives us some grounds for the belief that if we persist intelligently and doggedly in our bombing offensive we can rely to a considerable degree on the results of that offensive in itself to afford us protection from enemy bombing. But this will need to be supplemented by other means.

6. When the enemy bomber aerodromes are within fighter range—and they sometimes may not be—a mere offensive patrol of fighters 'trailing its coat' overhead is not likely to impel a wise bomber commander to come up and fight them; it is not his job, and he has a great many better things to do with his aircraft than indulging in gratuitous dog-fights if he

can avoid them. Low-flying attack with machine-guns and the
small bombs that fighters can carry is, however, likely to achieve
important results. Modern conditions, absence of hangars, and
due attention to dispersion may obviate serious actual damage
to personnel and material; but the effect on morale at least
is bound to be bad. So low-flying fighter attack on enemy
bomber aerodromes will often be worth while, and may at least
have the effect of causing the enemy to shift his bombers' bases
farther back and thus reduce their effective range. And it will
usually be necessary for fighter action to be supplemented by
bomber attacks on enemy aerodromes.

7. The Manual lays down that 'when conditions are favourable a temporary advantage in the struggle for air superiority
may be obtained by attacking enemy aerodromes'. There is a
tendency common in the Service to underrate the effect of such
attacks, which, if well executed under favourable conditions—
especially by bombers and fighters operating together—can
be well worth while. These conditions, broadly speaking, are
those of space and time—space, when aerodromes are few and
congested, and especially when the enemy air forces cannot be
rapidly reinforced; and time, when the inevitable disorganization—even if only temporary—may have serious results. In
point of material damage, reserve depots and parks may be
useful objectives, since there dispersion is less easy than on aerodromes of fighting units, and the resulting material damage
may be more widely felt.

8. Thus the best method of dealing with the enemy's bombers
—as with his fighters—will normally be by the maintenance
of an active offensive. On the other hand, it is dangerous to
make a fetish of any principle or to become the slave of any
tactical doctrine as the French did in 1914. Just as it may
sometimes be necessary—as previously explained—to divert
temporarily even the whole of our air forces to the *strategically*
defensive role for reasons of security,[1] so on occasions we
may be compelled for the same reason to divert part of
our fighter strength to the *tactical* defensive. This should be
avoided whenever possible; it should be the exception and
only temporary, and it can never guarantee complete success.
We in the British Service pin our faith to the offensive policy

[1] See p. 27 *ante.*

partly for its moral value and partly because as a rule it is only by taking the offensive that we can be tolerably certain of meeting and destroying the enemy in the air. So ordinarily our vital centres in the field must rely for their protection on the results of the air offensive, backed by such measure of direct passive defence by guns and searchlights as we are able to afford them. On the other hand, there may exist temporarily within air range of the enemy some point or centre so exceptionally vital to us that not only is the enemy morally bound to attack it (i.e. it becomes a certain 'point of enemy destination') but also that we cannot possibly accept the risk of failure to afford an adequate degree of protection by the maintenance of our air offensive alone. It will then become necessary to provide, at least temporarily, some measure of direct protection, to send a proportion of our fighters to meet the enemy bombers where they are certain to find them, i.e. in the vicinity of that vital centre. Two examples of such exceptionally vital points are the port of disembarkation while the army is landing in an allied country within enemy air range; and the transports, landing craft, and beaches of an army engaged in an opposed landing on an enemy coast. We could not possibly afford to expose either of them to the risk of unrestricted attack by such enemy air forces as might elude our offensive patrols over their aerodromes. And although any direct protection we could afford is bound to be uneconomical, is almost certain to be unsatisfactory, and cannot possibly guarantee complete immunity, it will fortunately—as in the examples quoted—generally only be temporary.

9. It has already been explained that one important method of forcing an enemy to action in the air is to attack something which is vital to him and which therefore he must defend. There is one situation in which this must be deliberately exploited as the *only* way of getting air superiority, and it is a situation of especial importance, namely the preliminary stage of an opposed landing on an enemy coast-line—a form of operation which has been a feature of almost every major war in British history. In the normal way a belligerent will not attempt to decline combat or to keep his bombers on the ground —for obvious reasons. And (again in the normal way) if he does so, then so much the better, we can get on with our air

offensive unhindered; an air force sitting on the ground can usually be disregarded, for, unlike the 'Fleet in being', it need exert no restrictive influence on our operations, nor does it constitute any threat. On the other hand, in the special circumstances prior to an opposed landing an air force 'in being' would constitute a very definite threat. It is a matter of common agreement in the three Services that to land troops packed in tows from crowded transports on open beaches is a sufficiently delicate operation under the most favourable conditions, but in the face of an unbeaten air force is not a feasible operation of war. And a defending commander, if he were able to maintain his air forces 'in being', either by keeping them out of range or by effective dispersion on the ground, might well make such a landing impracticable without firing a shot or dropping a bomb. Indeed, this situation is the one exception to the previously stated rule that the attainment of air superiority is not a preliminary phase to be gone through before the real business can begin: it is the one occasion on which the enemy air force must be reduced before we can devote our air effort to the main object—which in this case is to assist the troops in getting ashore, and in consolidating their position once they are ashore.

Actually it is almost impossible to imagine any situation of this sort in which the defending commander would in fact be able to afford to keep his air forces sitting on the ground inactive. Direct attack on them at their bases may be impossible or ineffective; but they could almost certainly be brought to action by attacking something which would be vital to them, and in that way forcing them to fight in its defence. For instance, in the Dardanelles the Turks could not have afforded to remain inactive in the face of air action on a modern scale against their bases at Constantinople, the Kuleli Burgas bridge, or their communications with the Peninsula by sea and road—or, if they had done so, then the necessity for landing the army in the face of opposition might never have arisen, since the maintenance of the enemy troops on the Peninsula would have become impracticable. And this in future may be the answer—the defending commander will have to choose between giving battle in the air, and thus weakening his position on the day of the combined landing, or accepting the possibly fatal results of unrestricted, unopposed air action against his vitals.

Principles
#1-3

Now to examine some of these principles in rather more detail. The first, the principle of the active offensive, is one which had many critics in the last War. People in the back areas, on the lines of communication, and at the bases complained of being bombed and, quite naturally, often failed to understand the absence of our own aircraft over them. The fighting troops in the line constantly saw our army co-operation aircraft being shot down; gunners complained that their air observation was being interfered with, and were told that our fighters were away over the enemy lines engaged on offensive patrols—and they not unnaturally wondered. Even in the air force itself the personnel of army co-operation and bomber squadrons sometimes complained of the lack of direct protection; and squadron commanders were left to draw what consolation they could for the loss of valuable pilots and observers, from the lists (published in the official communiqué) of hostile aircraft shot down over their aerodromes by our offensive patrols. These things were the visible effects, and criticism on these lines was natural and inevitable: what the critics, of course, could not see was the result which would have followed if our Higher Command had yielded to the often very severe pressure and fallen back on a defensive attitude. It is true that the detailed execution of the offensive policy was often not on the soundest lines; and that it was often kept up at times when the results to be achieved were not worth the heavy cost involved. But there is not the slightest doubt that as a policy its soundness was incontestable. After all, the proof of the pudding was in the eating. It was true that German bombing in our back areas was a serious nuisance, especially the night bombing which caused more than 4,000 casualties in the last six months of the war; but it was never really anything more, and our lines of communication were not subjected to the constant strain which the activities of our bombers must have caused behind the enemy line in the same period. It is also true that the work of close co-operation was often interrupted, and the corps squadrons went through some very bad periods; but here again in the long run the results unquestionably justified the offensive policy. The testimony on this point of the enemy army commander in the Somme battle has been quoted in a previous chapter; and even in the autumn of 1916, when the enemy was numerically at least equal to us in the

Somme area, the actual work of reconnaissance and artillery observation done by the German air service stood—according to statistics compiled at the time—in the proportion of only 4 per cent. of that done by our own squadrons. Even in the spring of the following year, in the Arras Sector (of evil memory), where the German squadrons, reinforced and re-equipped, were superior to the British both numerically and in technical performance, the tradition of the corps squadrons that no essential work for the army must remain undone stood in the main unbroken.

Fortunately we have not to rely entirely on conjecture for the results of lapsing from an offensive to a defensive policy in the air. The results on the German side have already been dealt with in an earlier chapter. On the side of the Allies we have another example in the history of the air fighting during the defence of Verdun in 1916. At the beginning of that battle, in which, of course, the French army was on the defensive, their air force was constantly offensive, bombing and patrolling far over the German line. As a result the German command fell back on the disastrous close protective barrage system—the *Sperrefliegen* already described. When this system proved ineffective, as it was bound to do, the Germans took other steps, reorganizing their Fokker fighters into separate units and adopting more aggressive tactics—largely owing to the influence of Oswald Boelcke. As an immediate result the French co-operation aircraft began to suffer more casualties, and a cry for close protection went up, to which the French command was eventually induced to yield. The effect was instantaneous.

'From that moment the tables were turned. The French air service began to lose its superiority, the French airmen could do very little reconnaissance, and their artillery pilots were compelled to work from inside the French lines. The French did not long acquiesce in this state of affairs. Thanks largely to the efforts made by Commandant du Peuty, they went back to the offensive, regained superiority, and with it all the advantages which they had temporarily lost.'[1]

[1] Official History, *The War in the Air*, vol. ii, p. 166. This volume contains as Appendix IX a memorandum on the offensive policy in air fighting issued by Head-quarters R.F.C. in September 1916; also an interesting letter on the subject from Sir Douglas Haig to the War Office dated Sept. 29th, 1916 (p. 297).

Unfortunately after du Peuty's death the French did not always adhere to his offensive policy. Perhaps owing to their national temperament and to the presence of many politicians in the ranks of their conscript armies, their high command was continually yielding to the demands for close protection, and till the end of the War they were constantly returning to the system of defensive barrage patrols—always with disastrous results.

While, then, the policy of the offensive is unquestionably sound, its execution must be intelligently directed. In the last war it was certainly not always applied as effectively as it might have been. Offensive patrols were often sent off without sufficiently careful planning and co-ordination with other forms of air activity. And, from the sound basis that the actual area in which other classes of squadron were working was not generally the best place to send the fighters, there tended to grow up the quite unsound theory that it *never* was so. This idea was by no means universal, and there were many examples of excellent co-operation between fighters and other classes; but there were also many occasions on which, for instance, bomber squadron commanders asking for fighter support were told that 'there will be an offensive patrol operating in the area at the time'—a fact which often had little apparent effect in reducing the number of enemy fighters encountered. In particular there was a feeling among bomber and corps reconnaissance pilots that offensive patrols over enemy aerodromes tended to become too much a 'set piece'—that they did not by any means always attain their object of meeting and destroying enemy aircraft in the air, and thus affording indirect protection to the bombers and artillery aircraft. On the opening day of the Somme battle, for instance, the offensive patrols found little to do, and in fact only one of them encountered enemy aircraft at all;[1] meanwhile, however, the bombers attacking railway objectives such as the junctions at Busigny and St. Quentin encountered serious opposition and suffered heavy casualties, and the efficacy of their bombing was very much reduced. It is true that the fact of the enemy fighters

Commandant du Peuty was a great air commander, who was wastefully killed commanding an infantry regiment. The barracks of the French Air Force at Thionville to-day bear his name. He was closely associated with Sir Hugh Trenchard in the policy of the air offensive.

[1] Official History, *The War in the Air*, vol. ii, p. 217.

being in strength about these railway junctions meant that the contact patrols and artillery-observation aircraft on the battle-front were able to work comparatively unhampered, which was rightly regarded as being of the first importance at the time. But it was also important to impose the maximum possible damage on Busigny and St. Quentin. And in order to achieve both these ends it was desirable to find and destroy as many enemy fighters as possible. If our offensive patrols sent to the enemy aerodromes at the time when they were most likely to find enemy aircraft there (whenever that may have been) found on arrival that the enemy was not there, then there seems no reason why they should not have had orders to proceed at once to the next most probable place—the enemy's likely area of destination, either over the railway objectives or over the battle-front, which-ever was considered more probable. What presumably hap-pened on this occasion—what too often did happen—was that the patrol became a sort of offensive barrage patrol between the enemy aerodrome areas and his probable points of destination. This idea sounds attractive, but in actual practice as often as not it was quite ineffective; such patrols, especially in cloudy weather, encountered the enemy largely by luck, and with a certain amount of patience and low cunning were fairly easy to evade. There were occasions when they were successful, but this was usually when the important area was small and when there were a very large number of fighters available. One such occasion was General Plumer's offensive at Messines in July 1917, which provides an example of the fifth principle stated on p. 36. Messines was a predominantly artillery battle: the overriding consideration in the air was to obtain complete superiority over the battle-front so that the artillery-observation aircraft—which were bound to become the main objective for enemy fighters—could work unhindered. And so in this battle the offensive patrols formed a sort of barrage from the ground to about 15,000 feet, over the enemy balloon line, and the opera-tion was completely successful. But the enemy balloon line opposite Messines was a quite definite and very limited area in which enemy fighters were bound to be encountered; and except in these special circumstances a system of barrage patrols, even though over enemy territory, will usually be uneconomical and ineffective—in future more so than ever, owing to modern conditions, the increased ceiling of aircraft, and developments in cloud flying.

In fact, the closer an offensive patrol can approximate to a *close blockade* of enemy aerodromes the more effective will it be. Against a strong air force in a highly organized theatre of war where aerodrome facilities are ample, such blockade can obviously only be partial unless our fighters are in overwhelming strength, and in the ordinary way it must be combined with other methods of bringing an enemy to action. In a campaign of this description close blockade can be very effective when its aim is to neutralize the local air defences temporarily, to keep the enemy's heads down while some particular operation is being carried out in the vicinity. There was a good example of this on September 25th, 1916: an important railway junction near Douai was being attacked by British bombers, and during the attack the three enemy fighter aerodromes in the vicinity of Douai were blockaded by fighters and bombers working together, being bombarded with high explosive and shrouded with smoke from phosphorus bombs, which effectively kept the defending aircraft on the ground. Incidentally, this interesting method of dealing with enemy aerodromes does not appear to have been much used subsequently for some reason; and it does give rise to a disturbing reflection as to the possible effects if instead of—or in addition to—phosphorus smoke an unscrupulous enemy were to make use of gas.

But as a general rule the method of close blockade will obviously be more effective in a country where aerodromes are fewer than they were in France. And it is therefore not surprising to find that most examples of its use in the last War are to be found in the history of our campaigns in the East, particularly in Palestine. On the day of Allenby's attack at Megiddo the blockade of the Turkish aerodromes was close and complete, and our aeroplanes had the sky to themselves. But there were other instances long before this, such as that on June 26th, 1917, when the enemy aerodrome at Ramleh was successfully blockaded during a bomber raid on the head-quarters of the Fourth Turkish Army in Jerusalem. In fact there were present in this campaign the two conditions which make this method ideal, namely a scarcity of aerodromes and a considerable numerical superiority on our side.

To sum up, offensive patrols over enemy aerodromes are usually the best and surest way of bringing an enemy to action; but they must often be combined with other methods, and they must always be intelligently planned and co-ordinated with other forms of air activity.

IV

THE SUPPLEMENTARY OFFENSIVE (*contd.*)

Principle #4

TURNING now to the next method, the co-operation of fighter squadrons in the bomber offensive. In the British Service a basic principle of bomber training is that squadrons must be capable of looking after themselves by close and steady formation flying, and the resultant mutually supporting fire from the rear guns. We have never favoured a policy of close escort for bomber formations, for usually very good reasons. In the last war the very word 'escort' was officially frowned upon, although in point of fact close escorts were often employed, and were very effective especially when composed of two-seater fighters, like the old F.E. or the Bristol Fighter. And we must beware of making a fetish of what is really a sound general rule; it would be thoroughly unwise to get into the habit of always sending fighters in company with bomber formations, but on the other hand close escorts are by no means necessarily vicious, and in fact are often essential. Each case must be examined on its merits and decided in the light of common sense. What is the problem? The primary arm of the air force is the bomber, and the aim of the bomber is to arrive over his objective and drop his bombs *accurately*. There are a variety of factors which tend to disturb this accuracy, such as anti-aircraft artillery or unfavourable weather conditions; but there is nothing more distracting to the bomber pilot than to find himself let in for a dog-fight over his objective, when all his energies and attention should be concentrated on that accurate use of the instruments which alone enables him to deliver his bombs on or near their target. So it is while he is *actually over* his objective that the bomber primarily requires freedom of action and protection from enemy fighter interference; and if our bombing offensive is being soundly directed and well executed, if the objectives selected are really vital to the enemy, it is precisely at that point that enemy fighters are most likely to be encountered. Therefore, remembering that the aim of the fighter offensive patrols is to find and destroy enemy aircraft, is not this an obvious place to which to send them? Now apart from other reasons which have been discussed in the foregoing pages, there are two main difficulties in the way of this course. In the last War—as to-day —the vast majority of the fighters were of the single-seater class, which has no rear gun; therefore when a bomber formation set out accompanied by fighters and encountered enemy aircraft on the outward journey, the escort had to turn in order to fight. The bombers obviously could not wait about till the issue was decided, and had therefore to press on without their escort to their objective—where they very often found more enemy fighters awaiting them. Meanwhile the fighter escort, even if the issue of the fight could be quickly decided, had not—and would not have to-day—a sufficient margin of speed over the bombers to enable them to catch up in time.

The second difficulty of course was—and for the moment still is—that the single-seater fighter has a relatively short endurance, with the result that the bombers' objectives were often out of fighter range. With the greatly increased ranges that may be expected within a few years it seems inconceivable that in a land campaign any of the enemy's vital centres, in the narrow sense of the term defined on p. 16, can be out of fighter range; that is to say beyond a range which fighters can attain, and attain with sufficient fuel in hand to allow of them having their fight and getting back again with a reasonable margin. So it will be seen that, as long as all our fighters are of the single-seater type, the problem resolves itself into a choice between two alternatives. The fighters can achieve their aim of meeting enemy aircraft in the air either by accompanying the bombers, with the chance of having to leave them before they reach their objective in order to engage enemy fighters encountered *en route*; or by meeting the bombers over their objective and engaging the defensive fighters at that point. And the decision should be based solely on the double test of which course is most likely to ensure that the bombs fall as near as possible to the target, not only on the actual raid in progress but on future occasions—and this in the long run depends mainly on the question of which course holds out the best chance of meeting and destroying enemy fighters in the air; since he who fights and is precluded for the best of reasons from running away, does not live to fight another day!

It has already been said that it is actually over the objective that the bomber most needs to be free from distractions. It is true that (if an Irishism may be forgiven) it is no good affording the bomber protection over his objective if he is shot down before he reaches it. But on the outward journey the bombers are not so likely to meet opposition as actually over their objective, and furthermore they are much more capable of looking after themselves. A well-trained bomber squadron flying in tactical formation is a very formidable object for fighter attack, it is not so likely to be forced to wide dispersion by anti-aircraft fire, and there is not the necessity for absolutely accurate and level flying that is essential during the actual attack on the objective. Moreover, the science of blind flying is making great strides to-day, and as this develops hand-in-hand with improvements in the science of navigation it will become easier for bomber squadrons on their outward journeys to rely for protection upon evasion, even to the extent of flying the whole way to their objectives actually in the clouds and only emerging over the target to drop their bombs. This increased capacity for evasion will have the result that interception by any form of defensive barrage patrol is even more ineffective than it has been in the past; and, therefore, that if an air force is compelled to resort to a tactical defensive the defending fighters will have no alternative to really close defence with the object of—at best—getting contact with the attackers over their objective, and at the worst of causing them casualties on their return journey and thus deterring them from further efforts. It thus becomes obvious that the attacker's fighters have the best chance of getting their fight in the immediate vicinity of the bomber's objectives; and with this in view fighter offensive patrols should be timed to arrive at those points a few minutes before the arrival of the bombers, so as to draw up the defending fighters and hold the ring while the raid is taking place.

So much for the condition when the bombers' objectives are within range of single-seater fighters. It is, however, in the nature of things that the single-seater fighter, as long as it exists as a class, will never have the same endurance as the bomber; and thus, even in a land campaign, there may be some objectives, less intimately but no less vitally connected with the maintenance of an army in the field, which are beyond the range of single-seaters. And therefore, as long as all our fighters are of this class, there will be occasions when close fighter support to bombers over their objective on the lines just described will be impossible. When this is so the bombers may have to be content with the indirect support of fighter offensive patrols over the enemy aerodromes; but there are other methods. For reasons just described, fighters are not likely to get contact with the enemy by flying in company with outgoing bomber squadrons as far as their fuel endurance will permit—though in certain exceptional circumstances, especially in very clear weather, this method need not be altogether ruled out. On the return journey, however, conditions are different, and the operation colloquially known as 'scooping out' may be very effective and afford invaluable relief and support to the returning bombers. There is often a critical period when bombers returning from a raid, having had a running fight all the way back from their objectives, and perhaps having suffered several casualties, find themselves beginning to run short of ammunition in the back guns, the level in the fuel-tanks getting too low to allow of their turning and fighting with the forward guns, and forty or fifty miles to go before regaining the shelter of their own territory. This experience was not unknown to the bomber squadrons of the Independent Force in 1918: many of the objectives in the industrial Rhineland were at very long ranges for the comparatively slow bombing aircraft of the day, and the German fighter pilots—though seldom so ready as our own to come in to close quarters—were adepts at long-range shooting with the front guns. Many a hard-pressed day-bomber formation in the last six months of the War, returning from Köln or Mannheim with several of their number missing, and perhaps a proportion of dead gunners or wounded pilots among the remainder, saw with relief a scooping-out patrol of British or French fighters coming to meet them fifty miles from home.

It will be noted that the main difficulties in the way of really effective co-operation between fighters and bombers arise from the fact that the present fighter is a single-seater with no rear gun and a limited radius of action. This class has another serious disadvantage for use in offensive patrols over enemy aerodromes. As the hostile vital centres are forced farther back, and as the necessity for the defence of those centres to be even closer than in the past increases for the reasons already

E

described, so an increasing proportion of enemy aerodromes will at least be so far back as to involve a long flight home over hostile territory on the termination of an offensive patrol. Twelve years ago in a memorial essay the present writer put up a plea for the inclusion in the home defence force of a two-seater fighter class for use mainly in offensive patrols over enemy aerodromes. 'The necessity for offensive fighters to be two-seaters is occasioned by the fact that on the termination of their offensive duties they will have to return over long distances of hostile territory, during which they will have to rely for defence upon close formation and covering fire from back guns.'[1] The fighter contingent with an army in the field will have to be drawn mainly from the home defence force; and unfortunately we have in the defence of London a special commitment, for which a very high rate of climb—in which, of course, the aeroplane carrying only one man is bound to be superior to that carrying two—is an essential requirement. For almost any other purpose, however, the two-seater is a superior fighting machine to the single-seater. Many a bomber pilot will have grateful memories of the self-sacrificing gallantry of the several famous squadrons which were armed with the old F.E., that good old general purpose day-bomber-fighter of 1916 and 1917. And statistics prove that, reckoning by the average numbers of enemy aircraft destroyed by different classes of squadron, the Bristol two-seater was by a considerable margin the most formidable fighter in the Service in 1918. Furthermore, the same tables show that the proportion of casualties suffered by squadrons of this class was appreciably lower than among the single-seater squadrons, for the excellent reason that the two-seater has a sting in his tail, and can defend himself while retreating. The fighter-bomber as a class is not being ignored in the British Service to-day; and one of the most important developments from the point of view of the air expeditionary force was the introduction in 1932 of the first squadron of two-seater fighters into the air defences of Great Britain. This squadron is equipped with an aeroplane, the Demon, of the same basic type as that with which the high performance day-bomber squadrons are equipped—and this is a very significant fact. In the field, the only means at present available of dealing

[1] *Gordon Shephard Memorial Essay*, 1923.

with enemy aircraft beyond single-seater range would be in effect to use a proportion of our bombing force as fighters. Whether they are used definitely as two-seater fighters, less their bombs but with the consequent improved performance and more ammunition, or whether they still carry their bombs and deal with enemy aircraft by bombing their aerodromes, is immaterial. The point is that they are being diverted from their proper task in the main air offensive to the subsidiary role of direct action against enemy aircraft. The answer surely is not to retain our present ratio of fast two-seater bombers and single-seater fighters, misemploying a proportion of the former when necessary, but the reverse—to include as high a proportion as possible of two-seater squadrons in our fighter contingent, so that we can use them as day-bombers if and when the air situation permits. We cannot afford to scrap the single-seater fighter altogether; we must always retain a few squadrons of a type which will be faster and have a better climb than the high performance day-bombers of any potential enemy. But these need only be relatively few, and the higher proportion of aircraft we can allot to the air expeditionary force that will be capable of taking their full share in the main air offensive, obviously the better. Therefore, while we cannot afford any modification in our home defence force that will adversely affect its efficiency in the primary role, the defence of this country, we can accept the advent of the first two-seater fighter squadrons in the home defences as a hopeful sign. And we must bear in mind, in framing our future programmes for re-equipment or expansion of the home defence force, that the more two-seater fighters it includes the better equipped it will be to meet its second great commitment, which is to constitute an imperial reserve from which can be drawn the squadrons of the air expeditionary force for an overseas campaign.

Occasions on which the single-seater fighter will remain the most suitable type are those when the employment of the fighters approximates to the tactical defensive, when the factor of interception requires a big margin of speed and a high rate of climb. Single-seaters will therefore perform more effectively the sort of duties outlined in the 8th principle on p. 37 above. It is unnecessary to elaborate this principle in greater detail, but there are two points in connexion with it which require emphasis. First it is as well to repeat that even local superiority can never be absolute; no defensive system, however strong, whether it is composed of short-range offensive barrage patrols as at Messines, or of definite local defensive patrols, can ensure that no determined enemy will get through. The air force commander who gives any guarantee of immunity should, therefore, be regarded with suspicion. The second point concerns the definitely defensive system, such as that designed to cover the disembarkation of an expeditionary force, and is that no system of defensive patrols or interceptors can possibly afford even a reasonable degree of protection without some system of warning, and without the co-operation of search-lights at night. It is obviously out of the question to expect in the field a system of air defence intelligence comparable to that in the home air defence organization—at any rate in the earlier stages of a campaign. Even in a highly civilized allied country like France, the difficulties of improvising the necessary system of communications upon the basis of a foreign civil telephone service would be insuperable. Moreover, actually in the defence of a base port there is the added difficulty that the wise enemy bomber will obviously make a detour and come in from the seaward side, where not only warning but also any adequate defence lighting is impracticable. Any scheme involving standing patrols is out of the question owing to the great number of aircraft that it would entail; and so it should be understood and accepted that interception of enemy bombers before they reach their objectives will be largely a matter of luck, and the best we can hope for is to distract them while over their objective, and inflict upon them there and on their return journey such heavy losses as may deter them from coming again. Nevertheless, even *some* warning is better than none at all, and we should consider whether it might not be possible to include in our field force some sort of air-defence-intelligence warning unit. It might be possible to organize a unit, which need not contain very many personnel or be very expensive, on the lines of the Observer Corps at home, possibly from the Territorial Army but on a higher standard of readiness for war. A number of observer groups of this sort, equipped with wireless, might be very valuable and very greatly increase the efficacy of any defensive system, by providing at least some warning to enable the defending fighters to get into the air earlier than would otherwise be possible.

Principle #6

It is officially recognized that attacks on enemy aerodromes by bomber aircraft are a diversion which is only worth while when adequate results may be expected, and when there are no more vital objectives against which such attacks can be directed. This principle has been dealt with in some detail in the foregoing chapters and needs no more elaboration further than to repeat that—certainly in a land campaign—this diversion can only be justified if by undertaking it we increase the efficacy of our main air offensive, or afford to our own army a security which cannot otherwise be attained. It is probably vain to imagine that we shall ever be able to dispense with it altogether; actually, in 1918, out of 666 tons of bombs dropped by Trenchard's force 220 tons, or almost exactly one-third, were dropped on twenty-three different German aerodromes. This represents a very serious diversion, and it is to be hoped—and if we get an adequate force of two-seater fighters it may be believed—that such a high proportion will not often be necessary in the future. But it is quite impossible to attempt to lay down any general rule as to the proportion of bombing effort that may have to be diverted to attack on aerodromes. Obviously it must vary in every set of circumstances. The Independent Force had to meet vigorous opposition by a first-rate enemy in great strength; in the final stages of the Palestine campaign, on the other hand, we had to deal only with a weak and demoralized enemy, and action against his aerodromes was left entirely to the fighters. So it is only possible to say that the extent to which air bombardment of aerodromes will be necessary must depend upon a variety of factors such as the strength, morale, technical efficiency, and fighting value of the enemy, the distance of his aerodromes from our own, and the air situation generally.

On the other hand it is worth examining briefly whether in fact 'adequate results may be expected' from this form of action, and if so in what conditions. In the British Service there is discernible a tendency to underrate the possibly serious results of bombardment of aerodromes on a modern scale. It is claimed that casualties to aircraft and personnel can be reduced to a negligible quantity by proper dispersion on the ground, and that bomb-holes in the aerodrome itself can be rapidly filled in. There is much truth in this claim. The effects of increased bombing efficacy have been at least discounted by other modern developments such as metal construction and greater independence of hangars; although the ease with which it is possible to recondition an aerodrome which has been well pitted with bomb-holes can be exaggerated. It is not impossible that permanent air bases in the vicinity of European frontiers may be provided with underground bomb-proof accommodation for the aircraft.[1] Nevertheless, apart from the possible effects of gas attack, which must be left to the reader's imagination, it is very dangerous to underestimate the damage and dislocation which might be caused even to-day by skilfully executed bombardment of aerodromes, especially if it is combined with fighter attack. Probably the lack of anxiety with which it is commonly regarded in the Royal Air Force is due to our comparative immunity from it in the War, and it is more than likely that it is held in greater respect in the German Service, and even in the French. There was always a certain amount of sporadic bombing, mostly at night, on or in the vicinity of British aerodromes in France; but it was usually little more than a nuisance, and it was not often that our squadrons suffered as seriously as, for instance, did No. 48 at Bertangles on the 24th of August 1918, when three-quarters of the aircraft were destroyed and the squadron had to be taken out of the line for several weeks to refit. Actually in the last sixteen months of the War there were nine instances of attacks on aerodromes of fighting units, on anything like an intensive scale; and the following little table summarizes the results:

Total number of enemy aircraft engaged (mostly large
   twin-engined night bombers of the Gotha type)—
   approximately   74
Total weight of bombs dropped—approximately . . 40 tons
Material damage, aircraft destroyed .   .   .   29
               aircraft damaged .   .   .   48
               hangars destroyed or damaged. . 19
                 and a great deal of M.T. destroyed
                 or damaged

[1] Even in the last War the Germans had some bomb-proof shelters for aircraft at their Belgian coast bases.

Personnel casualties—killed   26

               wounded   82

Enemy casualties in aircraft   3

Actually 21 out of the 29 British aircraft destroyed were lost in two quite minor raids, whereas the most intensive attack—that on Coudekerque on the night of June 6th–7th, 1918—in which 24 tons of bombs were dropped, only destroyed some hangars, and none of the aircraft were lost, because they were all in the air at the time; which shows how largely the element of luck enters into this sort of operation, particularly at night. Nevertheless the average works out at about one British aeroplane destroyed or damaged for every one German aeroplane engaged, and about eight and a half for every raid. Now this scale of loss when it occurs only about once every two months is not very serious, and in an air force about 1,500 strong, with a magnificent service of replacement behind it, is quite negligible. But if it were incurred by a force of only about one-third of that strength, with nothing like the facilities for rapid replacement of casualties, and incurred not once in two months but once in two weeks or still more in two days, it would become very serious indeed. So it is worth investigating whether anything like this scale of casualties is likely to result from the bombardment of aerodromes in the future, or whether altered conditions are likely materially to reduce it. This is a question of fundamental importance in air warfare, and it is essential to be quite clear about it, to understand first what were the conditions in which this scale of loss was incurred, and then to try to make up our minds to what extent modern circumstances have altered cases.

The first and most striking point is that in every one of the raids on the fighting aerodromes in the above table, as well as in those on the aircraft depots mentioned below, the bulk of the damage was caused by *hangars* being wrecked and usually burnt out. In the last War all our squadrons were housed in conspicuous canvas hangars, either of the small single R.E. or R.A.F. tent type, or, worse still, in the large wooden-framed Bessoneaux holding six or eight aircraft or even more. Even the former were usually arranged in neat rows close together along the edge of the aerodrome with—at best—low sand-bag walls between them to reduce the splinter effect of bombs; and from the latter there was seldom sufficient warning to get the machines out in time; and once one aeroplane was hit and the petrol-tank had caught fire, the remainder were almost inevitably destroyed. Now in this one vitally important respect conditions have changed to an extent which has altered the whole face of the problem. Even in the last War we could have done without hangars to a much greater extent than we did. In 1923 the Bristol Fighters of No. 4 Squadron stood out in the open on the Gallipoli peninsula for nine months under all conditions of weather; and more recently the Junkers Company operated an air line in Persia without any hangars for three years, the weather conditions varying from several feet of snow on the high aerodromes in winter to 130° in the shade at Bushire in the summer. In fact, metal construction and other modern developments in design have to-day rendered aircraft quite independent of hangars. At the same time, in the R.A.F. at least, improvements in engines and the removal from the squadrons of responsibility for any but small running repairs have obviated the necessity for large workshops, and greatly reduced the amount of transport required with the squadrons. Two or three light portable shelters will still be required in which running repairs and maintenance can be done in bad weather, but the bulk of the aircraft can be provided with engine and cockpit covers and picketed out in the open—just as a motor-boat is left out moored to a buoy. And instead of being picketed close together in neat lines dressed by the right (as we too often see them on manœuvres in peace), the aircraft must be scattered at irregular intervals all round the edges of the aerodrome. This is a nuisance, and will make defence, control and maintenance more difficult. But it is an inconvenience which can be minimized by training and practice, and in any case is an absolute essential; because by this means not only will squadrons become less conspicuous but also the material damage inflicted by air bombardment will be enormously reduced.

This brings us to the second point, which is that the more aeroplanes there are concentrated on any one aerodrome the higher proportion of casualties are likely to result from air bombardment; obviously the thicker the aeroplanes are on the ground the more likely are they to get hit. This factor, which really resolves itself into the question of the space available for aerodromes, is one which is likely to be aggravated by the need for wider dispersion of aircraft on the ground—though it may be partially alleviated by the fact that such fittings as wheel-brakes, wing-flaps, and variable pitch propellors will make it possible to use smaller spaces than were necessary fifteen years ago. Behind some portions of the British line in France, in Artois and Picardy, for instance, the country was almost ideally suited for aerodromes, and facilities were more or less unlimited; elsewhere, such as in the industrial north and in the dyke country of Flanders, it was very much harder to find big enough spaces for aerodromes. It may be a pure coincidence that 7 out of the 9 attacks on aerodromes referred to on p. 54 were in Flanders, but it is worth noting that on one aerodrome attacked there were about 70 aeroplanes, and on another about 50. The heaviest casualties on record were those suffered by the French air force during their concentration behind Verdun. In that area they had assembled no less than 630 aeroplanes on 7 aerodromes, all within a radius of about 3 miles, and on one aerodrome alone there were about 150 aeroplanes.[1] In the first attack on two of these aerodromes, which was at night, 60 aircraft were destroyed. It seems likely that prior to this the French had not taken all the steps that they might have to minimize the risk of casualties, because on the next night, when the Germans attacked 4 aerodromes containing in all 390 aircraft, only 25 were destroyed and 20 slightly damaged. The obvious remedy, of course, is never to have more than one or two squadrons on each aerodrome, and although this sounds rather like a counsel of perfection, it is the only solution to this difficulty. If we are up against a bold and efficient opponent in the air, and if the country in the immediate vicinity of the area of operations does not afford sufficient possible sites for aerodromes, then our air forces must be dispersed over a wider stretch of country, even at the expense of some sacrifice in effective range. This necessity, of course, raises a problem of control, in the solution of which there are two essentials. Firstly, an efficient field meteorological service, so that the Air Officer situated centrally can tell at any moment the weather conditions in any portion

[1] This is nearly as many as are concentrated at Hendon for the Annual R.A.F. Display.

of the area over which his command is dispersed, and if necessary adjust his programme accordingly. And secondly, a very comprehensive system of communications, including wireless, enciphering and deciphering machines, target codes, and the use of aircraft for intercommunication and personal visits between the A.O.C. and his subordinate commanders.

This then is a problem which, so far as it concerns the aerodromes of fighting units, can be solved—though we shall have to give it a great deal more attention than it is at present receiving in the British Service. Where aircraft parks and depots are concerned, the solution is rather more difficult. The number of reserve aircraft and the quantities of spare parts and stores maintained in the field will vary with the distance of the theatre of war from the sources of supply at home, but there will always have to be some. Even if the casualties due to enemy bombing of our forward aerodromes are reduced to a negligible minimum, it will still be serious if we cannot quickly replace normal wastage and battle casualties from our reserve parks and depots. Aircraft depots are organized and designed to serve a large number of squadrons, and there are consequently a large number of aircraft and a large volume of stores accommodated in them. It is difficult to understand why our great depots in the last War were not attacked more often, but when they were, the damage was very serious. Twenty-nine aeroplanes were destroyed in one raid on the depot at Dunkirk in October 1917; and at Marquise in the following September 13 German aircraft dropped about 12 tons of bombs which destroyed 27 aircraft and 25 engines, and seriously damaged 46 aircraft and many other stores—in each instance the bulk of the damage being done by hangars getting burnt out. The obvious solution to this problem is to establish aircraft depots out of enemy bombing range, and the disadvantages can be minimized by the use of transport aircraft for the carriage of engines and other spare parts—a service for which impressed commercial aircraft may be very suitable. But this solution will not always be practicable, and so we shall have to accept the obvious disadvantages and organize our maintenance and supply system in smaller units, and put fewer eggs into each basket.[1] Aircraft depots will in any event have to be situated on aerodromes, and if they

[1] See p. 118 below.

have to be in the forward areas, the reserve aircraft will have to be picketed out and dispersed on the ground in the same way as those of fighting units.

The third factor which we should consider in connexion with air bombardment of aerodromes is that of the facilities for replacement of casualties. In 1917 and 1918 the output of aeroplanes from industry was ample to meet all our requirements, although on the other side it was causing the Germans some anxiety at the end of the War. In the opening stages of a future war, however, it will be a very serious problem indeed; and until the normal peace production of the world's aircraft industries is very much larger than it is to-day, even those nations which are best situated in this respect will have great difficulty in finding sufficient aircraft to meet normal wastage and battle casualties in the first months of war. Therefore, when the other conditions are favourable, and especially when our Intelligence leads us to believe that the enemy is finding it peculiarly difficult to replace losses in his field air force from his aircraft industry, effective bombardment of his aerodromes may have very valuable results in reducing his air activity. And it is worth noting that the effect of such action against his aerodromes in the field may be rendered still more embarrassing by action against the centres of his aircraft industry itself, although this will not normally fall within the province of the air expeditionary force.

There is another factor, closely allied to the last, of especial concern to the British Empire, which has several isolated fortified bases of first-rate strategical importance disposed along the sea routes. Replacement of casualties may sometimes be impracticable, not because reserve or reinforcing aircraft do not exist, but because they cannot be made available where they are wanted in time. Even if the scene of operations is within flying range—and to-day at least one of our fortified bases is not—an air route is liable to be cut by the capture of even one essential refuelling aerodrome upon it. This situation, where a belligerent's air forces locally available cannot be reinforced, constitutes another condition favourable to the deliberate and systematic destruction by air bombardment of his aircraft on the ground; the more so because it is a situation in which the defending commander may endeavour, for reasons described on p. 39, to avoid premature loss by declining battle in the air. From the British point of view, of course, the obvious moral is the first-rate importance of as near as possible absolute security for the refuelling-grounds on the Empire air routes, and of the rapid development—by emergency tankage or other means—of longer cruising range for all classes of Service aircraft.

The last condition which need be referred to as affecting the question of the bombardment of aerodromes is one of *time*, and must be combined with others previously discussed if really adequate results are to be expected. A heavy bombardment of an aerodrome, even if it does not cause many actual casualties to aeroplanes, is bound to cause some temporary disorganization, and damage to the morale of the personnel; and a mass of bomb-holes in the aerodrome must cause some inconvenience and delay in getting large numbers of aircraft off the ground to a time programme. If, therefore, such an attack is launched immediately before some important operation, it may have valuable results in restricting and disorganizing the enemy's air activity at a critical time.

Finally, this problem of the air bombardment of aerodromes and bases has necessarily been discussed mainly in the light of our own experience as the object of such action seventeen years ago, and its importance under modern conditions in the scheme of air warfare as a whole must to some extent be a matter of conjecture. Although recent developments and the results of experience in the last War have in one direction undoubtedly minimized the results likely to be obtained, it must yet be remembered on the side of the attack that the performance of aircraft, the technique of bombing, and the destructive capacity of bombs have all made immense strides in the years that have passed since it was actually experienced. To sum up, it must be sufficient to say that the air bombardment of aerodromes is a form of air action which it is very dangerous to underrate, and which may, when the conditions are favourable, still achieve very adequate results.

PART II
THE SELECTION OF OBJECTIVES

V
STRATEGIC CONCENTRATION

'All bombing, even when carried out on very distant and apparently independent objectives, must be co-ordinated with the efforts that are being made by the land or sea forces, both as to the selection of objectives and as to the time at which the attacks shall take place. . . . It is utterly wrong and wasteful to look upon them as entirely separate duties.'[1]

IN the first part of this book we have considered the measures necessary to create and maintain an air situation sufficiently favourable to enable us to direct the efforts of the air striking force to the achievement of the object in a land campaign. And it seems advisable to begin the second part, which is an examination of some of the methods by which that object must be achieved, by a brief restatement of the object itself. The object of the air force in a campaign of the first magnitude in which great armies are engaged is the defeat of the enemy's forces in the field, and primarily of his army. In any future war, which even more than the last is bound to be a war more of material than of man-power, of machinery rather than of muscle, this object covers a field much wider than it is the purpose of this book to explore in any detail. And although this wider aspect of the problem is of such vital importance that it must later receive more than a passing reference, the object at least of those air forces directly co-operating with the army in a theatre of war can be reduced to more narrow limits—to operate in such a way as most effectively to contribute to the overthrow of the enemy army in the field.

This is not the place for a critical survey of the principles of war as codified and tabulated in the war manuals of the fighting services. Actually in the view of the present writer the whole matter has been overcodified, and the majority of the so-called

[1] Marshal of the Air Force Lord Trenchard, *Army Quarterly*, April 1921.

principles of war are not principles at all. But there are three great fundamental rules which are really worthy of the title of principles, and are described in the Field Service Regulations as the principles of *concentration*, of *offensive action*, and of *security*. These are the real principles, of which the observance is essential to victory, and to ignore which is to court defeat. And all the other factors described as principles are surely only elements in these three. Thus mobility, economy of force, and co-operation are elements without which concentration is impossible; surprise and again mobility are essential ingredients in a successful offensive; and it should be clear from the foregoing chapters that the correct application of the principle of security depends upon a balanced economy of force.[1] The principles of security and of offensive action have already received due recognition in the first four chapters of this book—indeed the main burden of those chapters was that air superiority, which is a measure of security necessary to ensure freedom of action, can only be secured by the offensive. And it is now necessary to consider the principle of concentration, which in air warfare, even more than on land, is the foundation and corner-stone of sound strategy. This principle is described in the Field Service Regulations in the following words:

> '*The principle of concentration*: The application of this principle consists in the concentration and employment of the maximum force, moral, physical, and material, at the decisive time and place (whether that place be a strategical theatre or a tactical objective).'

The tactical aspect of concentration is considered in a later chapter in the light of the actual employment of the air force at the battle of Amiens in August 1918; we are concerned here with concentration in the wider sense—with what may be described as the major strategy of air action in a land campaign.[2] Now

---

[1] See, for instance, p. 30 above.

[2] A condition which is often rather confusing, not only in air warfare, is the existence of what—for lack of a better expression—one may term *double* strategy and double tactics. Air warfare is not peculiar in this respect. This double nature of war has necessitated, long before the advent of aircraft, that complex range of definitions, from Imperial or Grand Strategy through that vague border-land where minor (or battle-field) strategy mingles with major tactics, to the genuinely minor tactics of the infantry section, for instance, or the individual warship. So in the air we have grand strategy as opposed to minor strategy—Chapter V of this book as against Chapter III; and major tactics as discussed in Chapter X as opposed

the capacity to concentrate the maximum force at the decisive time and place obviously involves as a first essential a clear understanding of *what is the decisive place at the time.* This is not always as easy as it sounds; the history of the last War is full of examples of serious disagreements on this point in high circles, from the 'Eastern' controversies of 1915 to the institution of the Independent Air Force described later in this chapter. But it must be assumed—and our improved defence organization gives us ground for hope—that it will be less difficult in the future. However that may be, some such agreement must be arrived at before it is possible to select in a general sense the objectives for air action. These objectives fall conveniently into two main classes, which may be briefly described as *fighting troops* and *supply.* Here it is necessary to elaborate a little. An army can be defeated by one of two main alternative means—not necessarily mutually exclusive: We can strike at the enemy's troops themselves, either by killing them or preventing them from being in the right place at the right time; or we can ruin their fighting efficiency by depriving them of their supplies of food and war material of all kinds on which they depend for existence as a fighting force. Thus under *fighting troops* as a general class of objective must be included not only the soldiers themselves but objectives such as the rail communications and roads on which they must depend for strategical and tactical mobility, or the head-quarters which control and direct their movements. The heading of *supply* covers a still wider field: it embraces the whole range of food-supply and munitionment, from the raw material in the mine through all the processes of production and manufacture, the depots at the base and on the lines of communication, right up to the first-line transport of the forward troops. But for the purpose of this examination it is convenient to consider supply under two sub-headings which we may describe rather arbitrarily as *Production* and *Supply in the Field.* If we consider under the heading of Production the provision and movement of food, clothing, weapons, ammunition and warlike stores of every description from their source up to their arrival in the area of operations; then the holding of reserves of such material in the theatre of war, and their

to minor tactics of air warfare in its most limited sense—aeroplane against aeroplane—which is outside the scope of this review.

distribution from the base depots throughout the lines of communication to the forward troops, fall conveniently under the heading of Supply in the Field. Actually within a theatre of war the distinction between objectives under the headings of supply and fighting troops is often nebulous; the closer those objectives are to the fighting zone the more nebulous does that distinction become, and action designed primarily to interfere with the one will inevitably affect the other. Indeed, the methods adopted will hardly differ and will ordinarily be narrowed down to attack on communications, which is dealt with in a later chapter. But between attack on production and fighting troops there will always be a very clear distinction, marked by the condition that the former will not, as a rule, be the responsibility of the air forces co-operating with the army in the field. There may be exceptions to this rule, but at least in a European war it is certain that operations against enemy production will be conducted by the Air Ministry, in accordance with a general plan approved by the War Cabinet on the advice of the three Chiefs of Staff—the more so because such operations will inevitably be connected with the problem of home defence. But the results of action of this nature may so vitally affect the operations of the field army that a brief digression to indicate its object and scope is relevant and necessary. In fact, a clear appreciation of the relation between attack on production and the more intimate form of co-operation in the theatre of land war is essential to a proper understanding of strategic air concentration—the fundamental principle on which is based the existence of a centralized autonomous air service. It is especially important that the soldier should understand this principle, and should realize how operations against enemy supply at the source may affect his own problems; because, as explained later in this chapter, there will be periods in a land campaign when it will be necessary in his own interest to withdraw at least some of the squadrons co-operating with him in the field, in order that they may be concentrated temporarily against enemy production.

This is not the place for an exhaustive treatise on air operations against industrial areas. Nor does the present writer intend to become involved in a discussion of such thorny subjects as the political, ethical, or legal aspects of the problem; the definition of military objectives; or the extent to which the exercise of direct pressure from the air on 'non-combatant' sections of an enemy community is justified, or expedient, or likely to be effective.[1] This last question in particular is one that has constantly arisen in connexion with air attack on production, ever since aircraft first dropped bombs beyond the actual zones of the armies in the last War. And since the end of that War it has assumed proportions probably in excess of its real importance, to an extent which has aroused all sorts of passions and prejudices, and has clouded the issue of how far, in fact, air action against war industries is likely to lead to the defeat of a belligerent, by depriving his armed forces of their essential munitions. It must not be imagined that the writer underrates the possibly terrible effects of air bombardment on the morale of a civil population. The importance of this factor has been recognized by most serious students of war, including Marshal Foch—whose well-known dictum on the moral effect of air action possibly becoming decisive has been quoted almost *ad nauseam* by writers on the subject. As a matter of fact it was as a measure of reprisal, an attempt to exercise moral pressure, rather than as an attempt to reduce the German output of munitions, that the first air operations on a serious scale against German industrial areas were initiated in October 1917. It is worth while asking ourselves—though we can never know the answer—to what extent the operations of the Independent Air Force in 1918 were responsible for the collapse of the will-to-resist in the German civil population, which was such an important contributory factor to the defeat of their armies in the field. In air operations against production the weight of attack will inevitably fall hardest upon a vitally important, and not by nature very amenable, section of the community—the industrial workers, whose morale and sticking-power cannot be expected to equal that of the disciplined soldier. And we should remember that if the moral effect of air bombardment was serious seventeen years ago, it will be immensely more so under modern conditions. This having been said, the subject will not be referred to again; but it should be borne in mind as a

[1] The reader who is interested in this very important aspect of the problem should read Mr. J. M. Spaight's *Air Power and the Cities*.

background to the consideration of the narrower—but probably not less important—aspect of the problem, the dislocation and restriction by air action of the supply of war material at the source.

This then is the *object* of attack on production, the dislocation and restriction of output from war industry, not primarily the material destruction of plant and stocks. Indeed, the material damage that can be inflicted by air action is as limited as its moral effect may be immense, measured in terms of disorganization and reduction of output. This is not to deny that under modern conditions the material damage and loss of life resulting from air action may both be very great, but only to emphasize that these are not the really important results, although, of course, the moral effect is in itself dependent, in the first instance, upon the potential material effects. Compared, however, with the absolute devastation and pulverization resulting from a first-class land bombardment, the material effect of air action will be very slight, while the cost of imposing it will be relatively almost negligible—a fact which the economic consequences of the last War may give us reason to regard as one of the compensating advantages for a new terror in warfare. When we speak of war industry it must be clear that we include not only the arsenals producing guns, ammunition, and tanks, the dockyards building ships of war, and the plant manufacturing aeroplanes. Without attempting to define military objectives—a task on which even the international jurists at The Hague in 1923 were unable to reach agreement—we can yet say with certainty that they will be held in future, as they were in the last War, to include the sources of production of those basic raw materials on which the true munition factories must depend for their supply. Thus iron and steel works, chemical works, blast-furnaces and coal-mines, electric-power stations and oil-wells, and the rail communications linking them together—all these things will undoubtedly be regarded as legitimate objectives in the campaign against production, and indeed all of them and more were attacked by one belligerent or another in 1918. This being so it is a very significant fact that of almost every first-class European nation, with the notable exception of the U.S.S.R., the great bulk of the war industries are concentrated in relatively small areas on or near the frontiers. Thus in France the great industrial areas of the Nord, of Briey, Lorraine, and Lyon–St. Étienne are all within air range of German or Italian soil; in Germany over 70 per cent. of her war industrial capacity is crowded into the small area of the Ruhr, while the mines and steel-works of Silesia are within air range of her eastern frontier. In Italy the percentage of industrial capacity within air range of frontiers is no less than 90 per cent., of which the bulk is concentrated in the Milan–Turin–Genoa triangle. In Poland the centres of the mining, metallurgical, and chemical industries, and in Czechoslovakia the bulk of the coal and iron-ore areas, are on or near the frontiers, within easy air range of potential enemies. In Great Britain we are little better off: the mining areas of Scotland and South Wales and the shipyards of the Clyde are, at least for the present, beyond the effective range of land-based heavier-than-air craft. But the bulk of our industrial areas are within range of modern bombers, operating from the Continent of Europe; and southern England, especially London, Woolwich Arsenal, the small-arms works at Enfield, and above all the London Docks, are within perilously close range. To us this means an added reason for our traditional policy of the maintenance of a weak Power in the Low Countries. For other European nations it raises a fresh and almost insoluble problem of air frontiers. It may turn out to be a factor making for peace, since nations may hesitate to embark on war with their vitals thus exposed to the new menace; but at present it seems more likely to amount to yet another irreconcilable conflict of security requirements, at least in northern Europe.

For evidence as to the *effects* of air action against industry we have to rely partly on the records of the Independent Force, but at present mainly on the recorded results of the German air operations against England. Figures are available showing the reduction of output of guns and ammunition from factories in the London area, the restriction of production of pig-iron in the Midlands, and the disorganization of rail traffic over wide areas of southern and central England. But for such details the reader must refer to the Official History, and to other less authoritative works on the same subject. There is only space here to state that a survey of the results of the German air raids between 1916 and 1918 clearly shows that even in those years they assumed very serious and alarming proportions. It must be borne in mind that in 4½ years well under 300 tons of bombs only were dropped in this country, a weight which at a very conservative estimate could be dropped in two or three days by any one of several first-class European air forces to-day. As the official historian reminds us,

> 'The air war was fought out on the Continent of Europe, and the bombing of Great Britain was episodic. It is not difficult to imagine circumstances in which we might have been called upon to meet the full force of Germany's air strength over this country Unless the reader ponders what that implies, the air raids will not be viewed in their proper perspective, nor will the potentialities of air attack be made clear.'[1]

It has been suggested, by a responsible and well-informed authority, that the results of air action on a modern scale in reducing the output from an industrial area might be tantamount to those of a hostile military occupation. The present writer has no wish to appear in the role of Fat Boy, and the above may be an overstatement. But it is difficult to resist at least the conclusion that air bombardment on anything approaching an intensive scale, if it can be maintained even at irregular intervals for any length of time, can to-day restrict the output from war industry to a degree which would make it quite impossible to meet the immense requirements of an army on the 1918 model, in weapons, ammunition, and warlike stores of almost every kind.

The *method* of attack on production again is a subject too big for detailed examination here. It demands a detailed and expert knowledge of the enemy's industrial system, of the communications linking the different parts of that system, and of the installations supplying it with power and light. It calls for the most careful selection of objectives, with a view to concentration on the minimum number; which in turn depends on a knowledge of the key components or basic products in the particular form of munition being attacked.[2] It involves the deliberate exploitation of the effects of air-raid precautions and warnings in reducing output, by interfering with work, and

---

[1] *The War in the Air*, vol. iii.

[2] For instance, iron-ore in the metal industry, or magnetos in the aircraft industry.

causing waste as well as actual damage to plant such as blast-furnaces—effects which can be spread over a wide area by the appropriate routeing of raids.[1] It may even involve the rejection of some excellent objective in favour of one, actually less vital, which can be subjected to intensive, as opposed to merely sporadic, attack. There is here no space for more than this brief indication of the scope and nature of a subject on which a book for itself could well be written, and of which at least the fundamentals were well understood by Foch and his great Staff Officer, Weygand, nearly twenty years ago.

Turning once more to the problem of strategic concentration: there were in the past—and to some extent still are—two opposing schools of thought on the subject of air attack on production, and its place in the nation's war plans. It would be a waste of time to enter into details of the sometimes fantastic claims of the extremists on either side, and it is only necessary to indicate here the views of reasonable and responsible authorities. Briefly, it was held on the one hand that the only way in which aircraft could contribute to a decision in war was by intimate and direct 'short-range' co-operation with the Army in the field, aimed at the traditional object of military overthrow of the enemy army. The late Sir Henry Wilson, for instance, as C.I.G.S. in 1918 and to the end of his days, was bitterly opposed to 'independent' air operations in any form.[2] The claim on the other side—submitted in the past, for instance, by Sir Frederick Sykes and to-day by Brig.-Gen. Groves[3]—is to the effect that 'the necessary uses of aircraft with the Army and Navy being ensured, any available margin of air power should be employed on an independent basis for definite, strategic purposes'.[4] For some obscure reason the designation 'strategic' is reserved by both these distinguished officers for operations against enemy towns or industrial areas, as opposed to what the latter rather scornfully describes as 'legitimate and obvious co-operation in

---

[1] For instance, the works at Bous in the Saar were only bombed seven times, but they received about 300 warnings, due to our bombers passing near by in passage to other objectives.

[2] Incidentally, one wonders how much harm was done by the constant use of this unfortunate and quite misleading word 'Independent'.

[3] See *Behind the Smoke Screen*, by Brig.-Gen. P. R. C. Groves.

[4] *Aviation in Peace and War*, by Maj.-Gen. Sir F. H. Sykes, p. 88.

the Battle'. Neither Sir Frederick Sykes nor General Groves, so far as the present writer is aware, has offered any solution to the problem of how, when, or by whom it is to be decided that the 'necessary uses of aircraft' for the other Services have been adequately ensured, and that a surplus is now available for other uses. It is not surprising to learn that it was found difficult 'to arrive at an agreement as to the minimum tactical and grand tactical requirements of the Army and Navy'[1]— and in point of fact no such final agreement ever was arrived at, or ever can be.

Both cases are in fact fundamentally and dangerously unsound. It is absurd to try to make out that air forces can never have any effect in contributing to a decision in a land campaign, except as a sort of super-long-range artillery on the battle-field. It is just as unwise to claim that the decisive point is never 'the Battle', or that there can be any rule, capable of final or universal application, whereby it can be decided that the requirements of the other Services in air co-operation have reached saturation-point. Consider what these claims amount to: on the one hand that air forces can and must only and always be used for one purpose; on the other, that it is under any circumstances justifiable to split the available air force into two distinct and more or less permanent divisions. Both claims show unmistakably a complete failure to appreciate the meaning or importance of the principle of concentration, or to understand the most fundamental and valuable quality of air power— mobility. The answer is, of course, that we must make full use of the mobility and tactical flexibility of air power; and concentrate the maximum force on whatever task is likely to be decisive, or to contribute most usefully to an ultimate decision, at the time. This is a principle of supreme importance in air warfare, and is in fact the key to the whole strategy of air power. And in order to emphasize and illustrate this principle it is worth undertaking a short survey of a most interesting phase in the last War, and to examine the circumstances attending the formation of the Independent Force in 1918.

From 1915 onwards there had been various attempts to undertake the methodical bombing of Germany. The only one which really materialized was the 3rd Wing of the Royal Naval Air

---

[1] *Aviation in Peace and War*, by Maj.-Gen. Sir F. H. Sykes, p. 88.

Service, which operated, sometimes in conjunction with the French, from Luxeuil for about twelve months, but was finally disbanded in June 1917. There were other proposals which came to nothing, such as the Detling detachment in the spring of 1916, and the abortive proposal to form a Franco-British bombing force in October 1916—both of which were stifled at birth by the opposition of G.H.Q. in France. In each of these cases the initiative was taken by the Admiralty—who should have been concentrating their attention on the provision of an adequate air arm for co-operation with the Grand Fleet. And all were principally notable for the absence of any apparent considered strategic policy underlying them, and for a complete lack of co-ordination with the general strategic situation at the time. The real nucleus of what afterwards became the Independent Force was the 41st Wing Royal Flying Corps, which was formed with three squadrons under G.H.Q. in France, and began operations from Ochey in October 1917. This wing owed its inception to a decision by the British Government to retaliate against the German night attacks on England, which at this time were causing serious interference with the output of munitions as well as affecting adversely the morale of certain sections of the public. It is noteworthy that the first properly organized attempt to bomb Germany on a serious scale was undertaken mainly as a reprisal for, and a deterrent to, enemy air attacks on England, and not primarily for its effect on German war industry. The 41st Wing was expanded by two additional squadrons and became the 8th Brigade in February 1918; and by the time the Royal Air Force was constituted on April 1st, it was engaged in bombing enemy industrial targets in Lorraine, the Saar, and the Rhineland. It is at this stage that the story becomes interesting and instructive from the point of view of its bearing on the problem of strategic air concentration. By this time the results of enemy air action against this country, and those which it was believed were being achieved by the operations of the 8th Brigade against Germany, had drawn attention to the 'long-term' strategic possibilities of the air bombardment of enemy munition centres. Actually this was no new idea. General Trenchard, for instance, nearly two years before had asked for a certain number of squadrons for fighting the German armies in France, and for others to attack the enemy in Germany. But the formation of the separate, autonomous Air Service created the first opportunity for the air authorities to put their ideas into practice, which they lost no time in doing. Sir Frederick Sykes tells us in his book that 'Fortunately in 1918 when I was Chief of the Air Staff, we managed to secure a margin[1] and formed the Independent Air Force in June of that year'. In following out the process by which this 'margin' was secured, it is important to bear in mind as a background the sequence of events on the battle-front at the time, and especially to note carefully the dates of the more important of those events, comparing them with those in the story of the negotiations in London and at Versailles.

On the western front Ludendorff was using the divisions released by the collapse of Russia in a series of desperate attempts to break the Allied line, and gain a decision before the British economic pressure became fatal, and before the American armies could have time to make their weight felt in France. When the Royal Air Force came into being on the 1st of April the great German drive towards Amiens had just been brought to a standstill. From the 9th to the 29th of April, when the 'St. George' offensive between Ypres and La Bassée was driving back the British Second Army almost to Haze- brouck, and Sir Douglas Haig was drafting his famous 'backs to the wall' order, the newly formed Air Staff were drawing up the plans for the formation of the Independent Air Force to bomb Germany. On the 27th of May, when General Sykes was 'securing his margin' and forming the Independent Force, the Germans launched another offensive, this time on a front of about thirty-five miles against the French Sixth Army on the Aisne—an offensive which was only finally stayed on the 6th of June, by which time the enemy was on the Marne. On the following day a further blow was launched, also against the French, between Noyon and Montdidier; and finally on the 15th of July the last German offensive of the War drove the French south across the Marne, where it was brought to a standstill with the assistance of British, Italian, and American reserves,

---

[1] It is only fair to point out that this conception did not originate with Sir F. Sykes. The idea of a surplus of aircraft over and above the requirements of the other Services underlay most of the measures leading up to the eventual creation of a separate ministry and service.

including the 9th (G.H.Q, Reserve) Brigade of the R.A.F. Air reconnaissance and other Intelligence sources provided early and unmistakable warning of these offensives, as well as of the next one which Ludendorff intended to—but actually never did—launch on the Lys on July 20th.

Meanwhile, what was happening in London? The reader must bear with a certain amount of detail, including dates, which are included because they are very relevant to the point at issue when compared with the story of what was going on at the battle-front and at G.H.Q. in France at the time. And once again it is desirable to emphasize that this story is not related with the object of scoring a point against any individual, or of attempting to minimize the very great services of certain distinguished officers. It is easy enough to sit back and criticize at leisure after the event; it is far harder to see clearly the right course among the crowding responsibilities and distractions of a great war. The following facts are set forth only to illustrate what the present writer believes to be the one supremely important principle in air warfare; in the hope that they may help commanders in the future to select that right course, in the event of our ever being faced with another great war in the future.

In May 1918 the Air Council proposed to the British War Cabinet the formation of an Independent Force to bomb Germany, to be under the command of Major-General Sir H. M. Trenchard, who, in his turn, was to be directly responsible to the Air Ministry. They suggested, in view of the fact that the formation of an Inter-Allied bombing force was already in the wind, that the broad lines of action should be laid down by the Supreme War Council at Versailles, acting of course on the advice of the military representatives. In other words, not only was the Force to be independent of the British Commander-in-Chief in France, but the Generalissimo, Marshal Foch, was to have nothing to do with it either—though why the Air Council should have imagined that this new proposal to make war by committee should be any more successful than previous attempts, is not apparent.

This latter proposal, however, had not received the sanction of the War Cabinet before the end of May when the Chief of the Air Staff, General Sykes, suddenly came out with a *fait accompli* in the form of an intimation that the British Air Council had already instituted an Independent Force to bomb Germany, and had obtained the consent of M. Clemenceau to the granting of the necessary facilities. There was a difference of opinion as to whether this policy should be extended by the creation of an Inter-Allied Force for the same purpose. Italy and the United States were in favour of the idea, but the French would not agree—understandably, in view of the fact that at that moment their Sixth Army was being pushed back, fighting desperately, to the Marne. It was even suggested that this force, if created, should not be subject to the Commander-in-Chief in the field but to the Supreme War Council; in this, however, Great Britain, by whom the suggestion was made, was in a minority of one, the other three Allies being unanimously of the opinion that it should be under the Commander-in-Chief.

Sir H. Trenchard took up his appointment as General Officer Commanding the Independent Force on June 6th—the day before the German offensive at Montdidier. He held very strongly that the time was now ripe to attack German industrial centres; and advocated that while maintaining the present relative strength in France we should steadily increase the numbers of squadrons engaged in the bombing of Germany.

Discussions on the proposal to form an Inter-Allied bombing force were continued in London and Paris for several months, and it was finally decided early in August to form an Inter-Allied Force to bomb Germany, under the command of the Generalissimo, as soon as the necessary resources in personnel and aircraft became available. Marshal Foch was accordingly told to draw up an outline plan for the employment of this force against military objectives in Germany. The necessary details were arranged between the French and British governments, and at the end of October General Trenchard was appointed Commander-in-Chief of the Inter-Allied Independent Air Force, under the supreme control of Marshal Foch. Actually, of course, the armistice was signed before the Inter-Allied Force had time to take shape, but the programme for 1919 was on an ambitious scale, and provided for the inclusion of strong contingents of bombers from all the principal Allies. It will be noticed that this story so far provides no clue as to how or from where this 'margin' that Sir F. Sykes speaks of—this surplus of aircraft over and above the requirements of the Army—was to be secured. The decision was that the Force was to be formed as soon as the necessary personnel and aircraft could be spared. But who was to decide when they could be spared? The Supreme War Council? the Generalissimo? the various Allied Governments or Commanders-in-Chief? Neither Sir F. Sykes himself, the prime mover in the scheme, nor Brig.-Gen. Groves, the head of the operations section at the new Air Ministry, have put forward any solution to this extremely difficult problem—unless it be by their actions at the time, from which it appears that as far as the British were concerned the deciding authority was the Air Council, even though its views might be in direct opposition to those of the General Staff at the War Office. Sir H. Trenchard's view was that aircraft could be spared to attack German war industry when the British and French air forces were each strong enough to defeat the German air force on the Western Front. He argued that the Allied air forces in combination would then have such numerical preponderance over the enemy as not only to give them real air superiority, but also to enable them materially to influence the battle on the ground. And he expressed his opinion that this situation had been attained by the end of June 1918. Now it is at least open to serious doubt whether at this time the condition postulated did in fact exist on the battle-front. But even assuming that the British and French air forces in June 1918 *were* each strong enough to hold and defeat the German aviation, it is nevertheless impossible to agree that this was any criterion. The problem was not one merely of holding and defeating the German aviation, but *all the German forces in the field.* It was not merely a question of *hitting* the Germans, whether in France or in Germany, but of hitting them *fatally and decisively*—and this we were unquestionably not strong enough to do on the battle-front at the time. The strange truth is that Sir H. Trenchard does not appear to have held a very high opinion of the influence which air action might exert on a land battle, in spite of our experience only three months earlier in the March retreat.[1] He admitted that he did not believe that bombing would turn a defeat into a victory or a

[1] See p. 105.

victory into a defeat; and produced the argument—a curious one for so forceful and determined a commander—that an increase in the air forces at the front over and above a certain unstated limit would merely result in each individual unit of the increased force doing considerably less work than the units of the smaller force had done. We can, of course, agree that bombing will not turn defeat into victory, or vice versa; but it is not a question of whether aircraft could turn defeat into victory, but of preventing defeat from developing into disaster; nor of turning victory into defeat, but of converting an inconclusive offensive into a real break-through. This may have been too much to hope for in 1918—though in the opinion of the present writer it was not, and the low-flying operations on March 25th and 26th on the Fifth Army front had already proved that air action could at least avert a local disaster—but it emphatically is possible to-day.

Returning to the contention that by the spring and early summer of 1918 we were strong enough in the air on the battle-front to justify the diversion of bombers against enemy war industry, it is worth examining briefly what was going on at British head-quarters in France at the time. Remember again the military situation on the battle-front. No sooner had one terrific German offensive been brought to a standstill on one part of the front, with a loss of hundreds of square miles of ground, than another was launched elsewhere; and the Allied armies were literally gasping for breath. Invalids, 'C 3 men', labour battalions, airmen of balloon sections, cooks and batmen were taking their places in the line, and the Allied resources in personnel and material were strained almost to breaking-point. Among these resources not the least formidable were the bombing aircraft, which in the opinion of many officers at the time were capable of making a contribution of the utmost importance to the defeat of a hostile offensive by interfering with the enemy's communications serving the front of attack.[1] Early in April, when the second German offensive on the Lys was in full swing, Major-General Sir Philip Nash, one of the

[1] Col. the Hon. M. A. Wingfield, who was on the Q branch of the Second Army Staff, was at this time conducting a private correspondence with Brig.-Gen. Philip Game, B.G.G.S., at R.A.F. H.Q., in which he urged the importance of properly directed railway bombing, in what then appeared to be rather exaggerated terms. See p. 112 below.

British experts in France concerned with questions of transportation, put forward some very valuable suggestions on the way in which the bombing of railways might contribute to the defeat of the German attacks. He suggested the general line of policy to be adopted, and the results which might be expected; he described the various vulnerable points on a railway and made recommendations, including certain objectives which he said were *not* worth attacking. He suggested the methods which from a technical point of view were likely to create most dislocation of traffic, and finally selected three sections of railway on which he claimed that almost complete interruption could be effected towards that sector of the front which, at that time, was the danger-point; namely, the line in front of Amiens where the German March offensive had just been brought to a standstill.

The Air Staff, to whom of course the matter was referred, did not attempt to contest the soundness of Nash's conclusions. They pointed out, however, that to subject the targets which he had selected to the necessary continuous bombardment would require a force of aircraft much larger than was at their disposal. (Note that General Nash had asked for only three stretches of railway-line to be cut, and that within a month of this date there were over 80 bombers in the 8th Brigade operating against Germany.) General Nash, however, was not to be lightly put off. He was convinced of the importance of the project, and pursued the subject with determination; with the result that ultimately a series of alternative schemes for railway bombing were prepared, to meet possible enemy offensives on different sectors of the front. These concentrated bombing schemes, as they were called, will be referred to in a later chapter. The point, however, which is of interest to the present issue is that the objections of the Air Staff to Nash's proposals were mainly on the grounds that they had not enough aircraft to implement them effectively. And it was constantly emphasized, in connexion with the concentrated bombing schemes, that the bombing force at the disposal of the R.A.F. Headquarters in the field was very limited, with the result that no really intensive bombing programme could be undertaken and only a very limited number of objectives could be dealt with. This was at a time when the military situation on the ground in

France was very critical; and yet it was being claimed elsewhere that we were strong enough in the air in France, and Sir F. Sykes was 'securing his margin' and forming the Independent Force to bomb Germany.

The real fact is, of course, that at this period of the War *there was no such margin.* Every available aeroplane should have been concentrated on the battle-front, which was unquestionably the decisive place at this period of the War; and the employment of the Independent Force against objectives in Germany *at this time* was an entirely unjustified diversion of effort. Let it at once be said that there is no suggestion that attack on production in general, or on German war industry in particular, was unsound. On the contrary there is not the least doubt that it was sound, that it should have been done much earlier than it was and on a much larger scale. But it should have been done only *at the proper time*, and the proper time was *not* when the Allied armies in France were fighting with their backs to the wall for their very existence. No reduction in the output from Lorraine or the Ruhr, no pressure on the nerves of the German industrial population, was of any avail if, meanwhile, the German armies were in Paris and the Channel Ports, winning the War in France. Perhaps the strangest feature in all this narrative is the apparently complete failure to appreciate the value and importance of that fundamental characteristic of air-power— mobility. These bombing squadrons are discussed almost as though they were fixed defences—so many required here, and when we have got them then we can put so many there. It is true that Sir F. Sykes says in his book: 'It was of course understood that in the event of either the British or French armies being hard put to it, the Independent Air Force could temporarily come to their direct assistance, and act in close co-operation with them.' And it is true that at the time the possibility of the participation of two or three squadrons of the Independent Force in the concentrated bombing schemes was discussed. But it never materialized. If the British and French armies were not 'hard put to it' in the spring and summer of 1918, it is devoutly to be hoped they never will be! Even so, this refers only to the possibility of using the Independent Force in defence. In the attack on August 8th, when the British armies returned at last to the offensive with the most ambitious objectives, there was no question of the Independent Force taking part. The unfortunately rare occasions on which it was used to co-operate in the attack on the ground, as for instance during the French and American offensives in the Argonne and St. Mihiel in September, apparently found no favour with Sir F. Sykes. In fact, in his book he comments adversely to the effect that 'theoretically machines of the Independent Force should not have been utilized for attacking purely military objectives in the Army Zone, such as aerodromes,[1] and their co-operation with the Army for this purpose shows that their true role was either not appreciated or not favoured by the French or other Commands'. An amazing theory.

When Marshal Foch was appointed Generalissimo at the crisis of the spring campaign, would it not have been possible to appoint—under his direction—Sir H. Trenchard as a supreme Air Commander, to act as air adviser to the Marshal, to co-ordinate Allied air policy, and to control an Inter-Allied Air Reserve? This Reserve might have included not only the 8th Brigade as it then existed, but the 9th G.H.Q. Brigade of the R.A.F., the French Division Aérienne, and the Italian Eastern Air Detachment. And it should have been used concentrated, *where its influence was most likely to be decisive at the time*—whether in defence against the great German offensive, or in attack, as at Amiens on August 8th, or against the enemy sources of production on German soil. This may have been an unattainable ideal at the time—very likely it was, though the germ of such a scheme was to be seen when the British Reserve Brigade went down to assist the French on the Marne in July.[2] But surely something on these lines should be the aim on which to direct our policy in the event of another great war.

So in a land campaign—except when some object other than the defeat of the enemy army in the field becomes temporarily of more urgent importance, as it has in the past and may in the future—the three main classes of objective are, broadly speaking, appropriate to three distinct conditions in the military

---

[1] For the fallacy of this argument in so far as it refers actually to aerodromes, see p. 7 *ante*.

[2] See also par. 3 of Foch's instructions to the Allied air forces—dated April 1st, 1918—App. XVIII to *The War in the Air*, vol. iv.

situation on the ground. Thus fighting troops are the primary objective during actual battle periods, whether in attack or defence. Attack on production assumes the first importance during what the Field Service Regulations describe as 'periods of comparative inactivity' on the ground. And midway between the two, supply in the field will probably be the most suitable objective during periods of preparation for battle on the ground, notably in defence when the requirements of secrecy and surprise need not be prejudiced by action to prevent the accumulation by the enemy of reserves of ammunition and other material necessary to the attack. This, of course, is not a hard and fast rule. For instance, the dislocation of his supply during a battle may be a very effective way of bringing an enemy's offensive to a standstill—in any event the measures taken to interfere with his fighting troops will in themselves to some extent dislocate his supply; while the movements of troops, the accumulation of reserves, and billeting or lying-up areas may often constitute very suitable objectives before the actual battle opens. But attack on production should definitely be reserved for periods when no very active operations on the ground are in progress, or known to be imminent. The claim has been advanced that air action against enemy munition production can and should take the place of old-fashioned and costly offensives on the ground. It is, of course, agreed that the old-fashioned offensive on the ground is definitely a thing of the past; and as air forces increase in strength and efficiency, and if permanent fortifications, field defences, and anti-tank devices endow the defence on the ground with still greater strength relative to the attack, it is arguable that no belligerent will undertake offensive operations on the ground at all. Armies may become mere holding forces to garrison frontier defences, from the cover of which air forces will attempt to reduce the enemy to impotence and ultimate capitulation by attacks on his essential services, and centres of war industry and transportation. That is as may be. If, however, an enemy does undertake a land invasion, perhaps by great armoured forces supported by air action, it is vain to pretend that the situation could then be dealt with entirely by air measures against the enemy country, or the invading army defeated by attacks on its supply at the source—because these measures would not have time to

Blitzkrieg

take effect. It must be assumed that in the opening phase of such a campaign the enemy would have reserves of tanks, ammunition, and other material, and will have taken steps for their protection. In the event of a deadlock on the ground, then attack on production would come into its own, and—coupled with other air measures against the enemy country—might well prove decisive. But even during a stalemate on the ground we cannot guarantee that air action against production will obviate all possibility of further offensive operations on the ground. It may have temporarily to be suspended, or it may be reduced in intensity, perhaps by a vigorous and resourceful defensive or by the incidence of some objective temporarily more important—such as the submarine campaign of 1918 quoted below; and if this happens, then one side or the other may gain a breathing-space in which to replenish and accumulate fresh reserves of material, rest his nerves, and generally fit himself to resume the offensive. Moreover, although it is true that material and machinery will play an even greater part in a future war than they did in the last, yet the production of that war material may—and very likely will—be considerably less vulnerable. Because, even in the sphere of material, quality will replace quantity, mind will replace mass. Consider, for instance, the vast industrial areas, coal-mines, iron and steel works, foundries, blast-furnaces, and factories covering hundreds of acres, necessary to produce the hundreds of thousands of tons of H.E. ammunition, the thousands of guns and motor-transport vehicles, the clothing, equipment, and small arms of an army on the 1918 model. Compared with these, the sources of raw material necessary for the production and the plant required for the manufacture—for instance—of aircraft and tanks, even in formidable quantities, present a much more difficult class of objective, of which the really vital centres will be more widely dispersed and more easily protected. A belligerent should be able to equip adequate modernized forces, and to carry on in the face of far more drastic restriction on output from his basic industries than would have been possible in 1918.

Nevertheless, although air action against production cannot yet altogether replace active operations on the ground, it can and should limit and reduce them. It should certainly replace the constant, costly, fruitless hammering against unbroken opposition which was such a disastrous feature of the last War; and, if properly applied, it should endow the offensive on the ground with much greater chances of success. Consider, for instance, the third battle of Ypres. The original attack in July may or may not have been necessary. But suppose that, once that offensive had been brought to a standstill, both the Allies had decided not to resume the attack, but to stand on the defensive on the ground, while they concentrated every available bomber throughout the autumn and winter against the industrial centres of Germany. If this had happened, would the enemy have been able to attack on March 21st, 1918? And how many thousands of lives, how many millions of shells should we have saved with which to meet that attack if it had materialized, or attack ourselves at a favourable moment if it had not? These are questions born of wisdom after the event, which obviously can never be answered. The moral is that just as in the narrow field of tactics, so in the wider sphere of strategy, action on the ground and in the air cannot be driven in blinkers, but must be complementary, and each must be applied in such a way as to make the best use of, and create the best opportunities for, the other. Among the many factors to be considered before undertaking an offensive on the ground, one of the most important must be its relation to the situation in the air. And the soldier must be prepared to postpone, or otherwise profoundly modify, his operations in order to get the best out of the air forces co-operating with him.

The title of this chapter is Strategic Concentration. It has been necessary in the foregoing pages to enter in some detail into the question of air attack on production, and its history in the last War, because a proper understanding of the relation in a land campaign between attack on production and the more direct form of co-operation with the army should make clear the one really fundamental principle in air strategy. The war aim of the air force is officially stated to be 'to break down the enemy's resistance . . . by attacks on objectives calculated to achieve this end'. The wording is left deliberately indefinite, because in every form of war, and in different stages of any war, 'the objectives calculated to achieve this end' will vary. The whole art of air warfare is first the capacity to select the correct objective *at the time*, namely that on which attack is likely to be decisive, or to contribute most effectively to an ultimate decision; and then to concentrate against it the maximum possible force, leaving only the essential minimum elsewhere for security—and possibly to contain superior enemy detachments. Thus in the ordinary course of a land campaign we have these three main classes of objective: fighting troops, production, and supply in the field, of which the first two at least are widely different in character. But it must not be supposed that even in a war where great armies are engaged one of these three will necessarily always be the primary objective of the air force. Even when we have an army in the field there may be periods —as there were in the last War—when the decisive task is temporarily not the defeat of the enemy army. For instance, although there were at the time British armies numbering millions in the field, the British Empire came perhaps nearest to defeat in the spring of 1917, when the one factor above all others that was very nearly decisive in favour of the Central Powers was the submarine campaign against British shipping. At this stage in the War, the correct strategic objective for the Allied air forces was surely the enemy submarine fleet and its bases, and the whole weight of the Allied air effort might have been concentrated to co-operate with the navy in dealing with the submarine menace. True, many of the German submarine bases were on the German coast, out of range of air action; but 30 per cent. of the enemy undersea fleet was based on Belgium, and even a relatively small reduction in the weight of submarine attack, for quite a short period in the early spring of 1917, might just have made the difference between defeat and victory for the Allies. It would surely have been worth accepting some risks and inconvenience on the battle-front, and even in the home defences, if, by doing so, the sky above the Belgian coast could have been swept clear of enemy aircraft, and the submarine bases on that coast subjected to heavy and continuous bombardment from the most effective heights. Is it too much to claim that under these conditions the Belgian submarine bases might have been rendered untenable? Perhaps it is, in view of the relative inadequacy of air bombardment in those days; but the example will serve its purpose as an illustration of air strategy—the concentrated application of the maximum possible force on whatever may be the decisive objective at the time. In the future, a similar situation may arise with hostile aircraft taking the place of submarines. It is more than likely that in another war a powerful enemy bombing force, based on Belgian aerodromes behind an established barrier of land forces, would constitute just as great a threat to the security of this country—and therefore also to the cause of our Allies—as did the German submarines in 1917. If this situation arose then the same principle would apply, and the whole power of the Allied air forces might have to be deflected temporarily from direct co-operation with the army to deal with the hostile air forces.

Finally, it must be obvious that the capacity to concentrate rapidly demands a high degree of strategical mobility. The aeroplane is, of course, a superlatively mobile instrument of war *within its own tactical radius*; but we cannot claim that our strategical mobility is, as yet, anything like as great as it should, and could, be. It is true that we have made great strides in the direction of long-range strategical mobility by the establishment of the Empire air routes—though even there theory outruns practice. But it is open to serious question whether our organization or administrative machinery is really founded on a sufficiently high estimate of the vital importance of mobility, or whether our training is actually fitting the Royal Air Force to be basically mobile. Remember that our air contingents for any overseas expedition must be drawn from the Air Defences of Great Britain—the Imperial Air Reserve. The primary responsibility of the Royal Air Force is, of course, the defence of this country against air attack. For this purpose it is essential that at least the fighter squadrons should be able to get into action with the promptitude and dispatch which we have been accustomed to associate with the London Fire Brigade. At the same time we should recognize that such accessories as tarmac aprons, fitted workshops, huge lighted hangars, and bulk petrol installations capable of refuelling whole squadrons in a few minutes, are the antithesis of true mobility—however essential they may be for the one specialized purpose. And we should be very careful not to become unduly dependent upon such static luxuries. Could any selected number of squadrons of each or any class operate at full capacity and at short notice from open fields, with the aircraft picketed out, and the personnel living and working in bivouacs and lorries? Is the system whereby the squadron is the administrative unit in the field really the one most suitable from the point of view of mobility? Or might a higher degree of elasticity and freedom of action for the fighting units result from a centralization of all administration, transport, and first-line maintenance in the Wing? Should we be satisfied with having no heavy bomber-transport units on the Home Establishment? The same criterion should be applied to our equipment. Are modern service aircraft really as independent of hangars and fitted workshops as—say —a destroyer is of dockyards? Can they be refuelled in the open in heavy rain without getting water in the tanks? Are our arrangements for bombing-up heavy aircraft as suited for use on uneven muddy fields as on smooth tarmac aprons? These are the sort of questions which we should continually be asking ourselves, for unless they can be answered satisfactorily, then our air force is not basically mobile. Although the unexpected usually happens in war, if there is one thing more certain than another it is that in war we shall have to improvise, and operate under conditions very different from those prevailing on service aerodromes in peace. And if we are not really basically mobile, then we are on dangerously wrong lines, for air strategy means concentration, and concentration depends upon mobility.

VI

FIGHTING TROOPS AND SUPPLY

THE examination in the last chapter of the subject of strategical air concentration, though necessarily brief, should serve to make clear the relation between what have been rather misleadingly termed 'independent' air operations, and operations against an enemy's forces in the field. With this necessary background the remainder of this book will be concerned only with the activities of an air expeditionary force engaged in close and direct co-operation with an army in a land campaign. It is in the nature of our imperial problem that British expeditionary forces may be required to fight in any one or more of a number of theatres of different natures and conditions of development, and against enemies of widely varying characteristics: yet it is possible to lay down certain broad rules, and to suggest a general line of conduct which will apply equally to them all. And although the following chapters refer mainly to the conditions of a war of the first magnitude, and are based largely on the experience of the last war in Europe, yet in wars of a different class (for instance, in a serious campaign across the north-west frontier of India, if such a thing can be imagined) the conditions will vary only in degree; and the aims and objectives of air action will be fundamentally the same, though their relative importance may differ from those of European warfare.

Turning now to the selection of objectives for air action in the field: it has already been said that there are two main classes of objective, fighting troops and supply—the latter being confined, as far as the rest of this book is concerned, to the supply in the field of food, clothing, and war material of all kinds from the base depots to the first-line transport echelons. Before going on to a more detailed examination of these main objectives and the possible effect upon them of air action, it seems necessary to refer very briefly to the *process* by which the actual targets for air action should be selected. This is not the place for a professional discussion on the system of command and staff duties in the field. It may well be that in the near future, as the size and relative importance of the air expeditionary force increases, we shall find one Commander-in-Chief may be a soldier or an airman, served by a small, sp selected, combined staff, in supreme command of two con gents of coequal status, the air and the land forces, each with its own subordinate commander. This is a question of real importance, and we should take care to secure an amicable and reasonable settlement, in the leisured atmosphere of peace, of any differences of opinion which may lead to rancour and misunderstanding between commanders in war. The almost disastrous after-effects of the Curragh incident, and the consequent relations between certain higher commanders in the B.E.F. of 1914, should serve as a warning in this respect. It is not an easy problem: the system of 'two Kings in Israel', of co-operation as opposed to command, is never very satisfactory. It is certain that in these days, when our air frontier is on the Rhine,[1] we cannot go back to the old idea that any and all air forces in the field should be automatically under the command of the General commanding a small land expeditionary force of four or five divisions. Nevertheless it is a problem for which common sense and our national capacity for compromise will find a satisfactory solution; and anyhow, as far as this book is concerned, the fundamentals of the problem are not affected. As at present laid down in the training manuals, the Air Officer Commanding the air expeditionary force has a dual capacity: he is the executive commander of the air forces in the field— other than those army co-operation squadrons under the orders of army corps, and for these also he is responsible in matters of administration and technique; he is also appointed to General Head-quarters as the adviser to the Commander-in-Chief on all matters concerning the air force. This is the first important point to notice—the Air Officer must be in close and constant touch with the military commander, and Air Head-quarters will always be in the immediate vicinity of G.H.Q., of which at present it is virtually an integral part. This being understood, it will be clear that the evolution of the air plan will in its initial stage be the result of discussion between the military and air force commanders. Certain important subsidiary considerations which the latter has to keep in mind, such as the proportion of his striking force which may have to

[1] *Vide* Mr. Baldwin's speech in the House of Commons, July 19th, 1934.

be diverted to the achievement and the maintenance of a favourable air situation, have been referred to in a previous chapter.[1] Subject always to such considerations, the air plan in outline will be threshed out over the map in discussion between the two commanders with their principal staff officers, and will crystallize in the form of a decision by the Commander-in-Chief as to the part to be played by the air force in the achievement of his object, whatever it may be at the time. This decision then becomes, in technical military jargon, the 'object' on which is based the subsequent detailed appreciation by the operations branches of the Air and General Staffs; from which in turn emerges the definite plan of air action, including the detailed objectives to be attacked.

The Training Regulations define an appreciation as 'a military review of the situation, based on all available information, culminating in a statement of the means recommended to meet it'. It is hardly necessary to labour the importance, in making this review, of an intelligent use of all the available *information* about the enemy. It is obvious that the selection of those objectives whose destruction or interruption will be most inconvenient to the enemy presupposes an intimate acquaintance with such details as the locations and condition of his forces, particularly of his reserves; the communications, road or railway, which he is likely to use, and their capacity and limitations; the location of his head-quarters, signal-centres, supply and ammunition depots and maintenance establishments; and in fact his whole system of organization, transportation, and supply in the field. Based on such knowledge the staff must try to put themselves into the enemy's mind, and deduce as nearly as possible his probable alternative courses of action. All this is part of the normal functions of the Intelligence branch of the staff, and the principal change since the coming of air action has been that the zone of which a detailed knowledge is essential has increased in depth. It has always been important to be able to deduce, for instance, that certain reserves can arrive on the battle-field by a certain time; to-day it is important to know not only where these reserves are likely to come *from*, but their exact locations before they move and the exact routes which they are likely to use to reach the battle-field.[2] Moreover,

[1] See p. 29. [2] See p. 177 below.

certain details which were formerly of only academic interest
are to-day of first-class importance, such as the exact buildings
which housed the main Turkish telephone exchange at El
Afuleh, or the office from which that capable official 'D. R. Ost'
controlled the vast network of intricate railway communications
on the German eastern front.[1]

But there is another aspect of information of which the
importance is not so commonly recognized. Operations staff
officers cannot themselves select off the Intelligence map the
most suitable objectives in an enemy's system of supply and
transportation. Detailed intelligence about the enemy must
be supplemented by *expert technical advice* from representatives of
our own supply and transport services, railway or dock operating staffs. For instance, in any railway system there are a number of technical factors such as gradients, watering facilities, or
signalling arrangements which affect the capacity or efficiency
of lines; and only the practical working experts can say with
anything like certainty what points in that system are most
vulnerable from a technical, operating point of view. It is
difficult to overstate the importance of this step in the selection
of objectives for air action in any but the most straight-
forward of circumstances. If the staff officer tries to do it by
the light of nature he will be right only by luck; whereas the
expert, who will always be available in the transportation
directorate in war, may be able to lay his finger on the most
apparently improbable objective which is really of first-class
importance.

Before leaving this part of the subject it is necessary to refer
briefly to *reconnaissance*. It is obvious that the accurate information on which the initial plan is based must itself have been
obtained largely by air reconnaissance and photography. And
good reconnaissance is no less essential after zero; objectives
may have to be changed, and changed quickly; earlier information may prove to be inaccurate or out of date, or assumptions
may be falsified. So reconnaissance must be continued, not only
to confirm existing information and hence the initial plan, but
also to keep the commander supplied with up-to-date information on which to modify that plan if necessary in course of
execution. This is almost a truism, but one which has been

[1] See pp. 97 and 210 below.

known to be overlooked.[1] Finally it is worth noting that an important supplement to reconnaissance is an adequate signal-organization by which its results may rapidly be disseminated, because during a battle the best opportunities for air action will often be fleeting.

FIGHTING TROOPS

In considering the generic objective of *fighting troops* it will be convenient to examine the subject under two heads: as an objective first for bombers operating on normal lines whether by day or night; and secondly for a specialized form of tactics which is reserved for occasions of great emergency, namely low-flying attack. And under both heads it is possible to formulate certain rough rules which can be applied with the necessary minor adaptations to any type of warfare.

1. The first general rule, which applies equally to bombers and to assault aircraft,[2] is that *the aeroplane is not a battle-field weapon*—the air striking force is not as a rule best employed in the actual zone in which the armies are in contact. There are, of course, exceptions to this rule. One such exception may be during the initial break-in to a highly organized defensive system.[3] Another may be when it is necessary to break up a counter-attack, such as when 80 French bombers and 40 fighters caught a German division massing in the valley of the Savières near Villers Cotterets in July 1918, and dispersed it; or when six aircraft of No. 1 Australian squadron, during the second battle of Gaza, found about 2,000 Turkish cavalry and 800 infantry massing for a counter-attack in a wadi near Hareira and broke them up. Aircraft will usually have to be directed, in the defence, against targets closer in than in the attack, for the obvious reason that their effect must usually make itself felt more quickly in defence than in attack. But the general rule holds good. In theory a sound economy of force should require that a highly mobile weapon like an aeroplane should not be used to engage targets which can equally well be taken on by

---

[1] See, for instance, p. 173 below for neglect of reconnaissance during the battle of Amiens, August 8th, 1918.

[2] There is no official designation for low-flying attack aircraft; but in order to avoid the constant use of this rather clumsy expression, aeroplanes employed on this duty will be referred to in this book from now on as 'assault aircraft' the definition of which is set out on p. 98 below. [3] See p. 101 below.

other less mobile weapons on the ground. And theory is supported by experience, which goes to show that usually far more profitable and important objectives will be found farther back—beyond the range of artillery and machine-guns. An analysis in a later chapter[1] of the employment of the air striking force in the battle of Amiens shows that, while the bombers were mainly engaged in attacking the Somme bridges, and the assault aircraft were employed in close co-operation with the advancing infantry and tanks all west of the Somme, to the east of that river the roads and railways were choked with the German reserve divisions which were allowed to arrive relatively unhindered on the battle-field, and brought the advance of Rawlinson's army to a standstill on the fourth day. Very much the same thing happened at the battle of Cambrai in November 1917. On November 22nd, for instance, the majority of the assault aircraft were engaged—very gallantly and at a terrible cost in casualties—in close co-operation with the infantry, attacking enemy machine-guns and forward troops in the desperate fighting about Bourlon Hill and the village of Fontaine. Meanwhile, behind the battle-field, the medium reconnaissance aircraft were reporting 'a congestion of trains in Douai station—much movement south from that railhead—columns of troops and transport marching on Cambrai from Douai—other columns moving south on the Lens–Douai road'—and other similar movements, most of which were allowed to continue entirely unhindered from the air. A German division *en route* to Italy was stopped and detrained at Cambrai, and altogether between November 20th and 29th 100 trains a day brought in thirteen reserve divisions and 600 other units—batteries of artillery, engineer companies, and so on—to the German Second Army front. It is true that a programme of railway bombing had been prepared, which could not be carried out owing to the foul weather which prevailed almost throughout the battle. But even so, better results would have been obtained, and at less cost, if the assault aircraft had been used farther back; and in any case the story will serve as an example of the reason for using the air striking force against the enemy's rear communications and reserves, rather than against his forward elements on the actual battle-field.

[1] Chapter IX.

2. It must be obvious that large *movements of troops by road* are very vulnerable from the air, and especially by assault aircraft: so much so that it appears at least open to doubt whether it will ever be practicable in the face of a strong air force to move by day the huge columns of infantry, horsed artillery, and transport that comprise the division as we know it to-day. Armoured forces are much less vulnerable—or anyway the actual fighting echelons. They can afford to ignore machine-gun fire from the air; and although a hit even from a small bomb would put a tank out of action and cause a serious block on a road, yet the cross-country performance of armoured fighting vehicles will often enable them to get off the road at the first threat of air attack. Defiles such as towns or bridges must be avoided as far as possible; where they cannot be avoided they will have to be picketed by mobile anti-aircraft artillery and machine-guns—which will at least reduce the accuracy and intensity of attack—and passed through as quickly as possible, when possible at night.

A serious disadvantage of armoured forces—which indeed applies almost equally to other arms in this era of mechanization of artillery and transport—is that in these days, when the air constitutes a threat to communications more formidable than anything hitherto dreamt of, mechanized forces are entirely dependent for their fighting efficiency upon uninterrupted communication to the rear. The days when great mobile forces —like Budienny's cavalry divisions in the Russo-Polish war— could live on the country for long periods, have gone for ever. In the amusing words of an American officer, 'without gasoline machines are junk'; and the huge bulk installations at the base and the great masses of petrol on wheels at any moment behind a modern army do present a very serious problem. Nevertheless it is one which has got to be faced: whether we like it or not, the horse for most purposes of serious warfare is as obsolete as the ass's jawbone, and the internal-combustion engine in the end will reign supreme. It is a problem of which the difficulties and dangers can be minimized. Indeed, in many possible theatres of war—for instance in the Middle East or central Asia—the mechanized force has a definite advantage in this respect; because it does not require anything like the quantities of water that are essential to the army on an animal basis, nor is its fuel so bulky as the forage required for the horse. Nevertheless petrol is a terribly vulnerable form of supply. The universal adoption for military vehicles of heavy oil fuel—already well advanced in commercial practice—will restrict the damage that can be inflicted by successful air attack. But this will not be a complete solution; and the anti-aircraft protection of fuel-supply echelons is a problem which must be seriously tackled.

In the attack on troops moving by road, whether they are armoured forces or the traditional arms, there is one condition which must be observed: the air striking force must have *room* in which to make their weight felt. The air arm cannot remain permanently in contact; it has no actual physical stopping-power comparable to that of machine-guns and wire on the ground; in fact it *cannot hold*. It must therefore be enabled to make its initial attacks far back, so that they can be repeated again and again before the enemy has time to reach his objectives and deploy. A force deployed is comparatively immune from air action; the same force in column of route on the road is extremely vulnerable; and provided it can first be attacked sufficiently far from its destination, it may very likely be prevented from reaching that destination at all.

3. The *movement of troops by rail* is if anything more susceptible to interference from the air than movement by march route. The problem of air action against rail communications is examined in some detail in the next chapter; in summary, the technique is different from that of action against road movement, and depends as a rule for its effect more on interference with movement by the dislocation and disorganization of the railway system than on material damage or casualties inflicted on the troops themselves. Actual examples of this in war are scarce, because, for reasons which are explained elsewhere, railway bombing was never really adequately exploited in France. We have to rely mainly on our imagination to assess its probable results, assisted by negative examples of what might have been —such as those of the battles of Cambrai and Amiens already referred to. We know that the enemy bombing of Châlons and Chantilly in March 1918 did impose the relatively slight delay of ten hours on a number of French divisions, which were being rushed in by rail to help stem the German advance on Amiens— and even ten hours might conceivably be decisive in certain

circumstances. It is probable that more useful lessons could be derived from the German records of the results of our bombing of Thionville and Metz Sablon referred to in a later chapter. Obviously, however, it would be very unwise to deduce from the paucity of its achievements nearly twenty years ago, that railway bombing in modern conditions is the least likely to be equally negligible.

It would be equally unsafe to underrate the possibly disastrous results in the form of actual material damage, and casualties to personnel. Troops moving by rail, and especially when actually entraining or detraining, constitute a very concentrated and vulnerable target. On the opening day of our attack on the Somme, July 1st, 1916, the German 71st Reserve Regiment was being entrained at St. Quentin to move up and reinforce the defence. The three battalions of the regiment, mostly with arms piled and equipment off, were caught by British bombers when actually engaged in entrainment. A direct hit on one of the station buildings which contained ammunition caused an explosion which immediately spread to an ammunition train standing in a siding. The resultant fires and explosions destroyed the troop trains containing the equipment of two battalions; and, to cut a long story short, instead of going up to the Somme front the regiment was marched into billets in Ham to refit and recuperate, leaving 180 casualties in the station, and with its morale seriously shaken. This incident will serve to emphasize the point, already made with reference to bombing of road movement, that the initial attacks should be made *as far back as possible*—consistent with the capacity to execute them really effectively, and with economy of force in the selection of as few as possible really vital rail objectives.[1] From which it follows that, when there is a choice of attacking either entraining or detraining stations, the objective selected should —other things being equal—be the former.

4. It is the writer's belief—and profound regret—that in the British army at least the day of the horse in war has passed, for reasons quite unconnected with air action. The aeroplane has surely sealed the fate of an animal which did (whatever the pacifists may say) lend some pleasure, some romance, to an otherwise dirty business. *Mounted troops and animal transport* are

---

[1] See p. 178 below for an example of this in the Amiens battle.

indeed especially vulnerable to air action, and have the added disadvantage that horse or camel lines are usually almost impossible to conceal from the air. Compare[1] the 800 horses and 70 wagons of the present Field Brigade R.A. with the 100 motor vehicles of its mechanized equivalent; or the 570 horses and 36 motor transport of the cavalry regiment with the 38 armoured cars and 40 other vehicles of the armoured-car regiment. In June 1916 one high-flying bomber stampeded the horses of two regiments of an Australian light horse brigade near Romani—horses were being recovered from all up and down the Canal for weeks afterwards. And a field artillery brigade, caught in bivouac one evening by an enterprising German fighter pilot, lost 25 horses killed in one battery near Bourlon in the autumn of 1918. These two examples surely indicate that air action on a serious scale could at least temporarily immobilize horse-drawn artillery or transport; and its effects can be imagined, for instance, on the enormous mass of camel transport that was required to maintain the army in Palestine. It seems certain that the days of cavalry are numbered, except possibly a few units for reconnaissance; the French have for some years been converting their cavalry into mechanized units and 'dragons portés', and even in the British cavalry the armoured car, the light tank, and the motor-lorry are gradually taking the place of the horse.[2] Nevertheless we may yet have to deal with mounted troops in war—especially in certain areas of eastern Europe or central Asia. And if we do, it is worth remembering that the air machine-gun, though alarming, is actually a surprisingly innocuous weapon. It will become even more so as the increase in the speed of the aeroplane increases still further the intervals between the strike of the bullets. The weapon to use every time is the small bomb, which not only has great destructive effect in itself, but also by its noise is very liable to stampede the animals against which it is used.

5. *Troops in rest or reserve* as an air objective. It is by now a common-place that the strain exerted by modern war upon the soldier's nerves is even more harassing than that exerted on his

[1] All these figures are approximate, and exclude guns and limbers and motorcycles.

[2] This was written in 1934. The reorganization of the Army initiated early in 1936 will go a long way towards the total elimination of the horse in the British Service.

muscles; and it is not too much to say that occasional periods of real rest and nervous relaxation are essential if he is to maintain his morale and fighting efficiency. The elastic of his nervous system can be stretched almost to breaking-point very often— but provided it is periodically relaxed it will not break. Uninterrupted tension, however, will snap the strongest nerves. In the last War, though bombing on both sides was often very irritating, it is generally true to say that once a soldier was back out of range of hostile artillery, he was able to enjoy considerable periods of relaxation: he may have—and usually did—have plenty of hard manual labour, route-marching, training, and physical exertion of all sorts; but he was able to feel that he was temporarily out of danger. He came back into rest billets; he went to cinemas and concerts in the evenings; he went to various courses of instruction at Corps or Army Schools; occasionally (very occasionally!) he went on leave, and crossed the Channel —or went back by train to Berlin or Paris or Vienna or Moscow—in discomfort certainly, but in safety; he passed through staging camps and rest camps, travelled in troop trains or 'hopped his lorry'—never comfortable, often extremely weary, but for the time being secure from the imminent danger of being mutilated or hurled to sudden death. But those days are gone for ever. In future the soldier will never feel safe anywhere within two or three hundred miles of the enemy. The bomber, and particularly the night-bomber, will banish sleep, and in billets, rest camps, schools, and cinemas will engender a feeling of permanent insecurity that cannot fail to have a terribly wearing effect on morale. Nor will the effect be felt only in the vicinity of the objectives. Careful routeing of raids over areas known to be occupied by troops will spread the moral effect; and by constant alarums and excursions, dowsing of lights, and expeditions to bomb-proof shelters, will cause continual interruption of rest and relaxation that will be little less irritating for being unaccompanied by more lethal results. That the actual casualties caused by attack on troops in the back areas may be very serious (and after all the moral effect depends in the first instance upon material damage) is amply indicated by the results of the raid on the night 19th–20th of May, 1918, when eleven German aircraft caused no less than 840 casualties in the crowded rest camps and hospitals among the sand-dunes about Étaples.

6. No one who has had experience at any considerable _head-quarters_—even if only in peace-time exercises—of the accurate and intricate staff work that has to be done, always against time, in drafting the plans, co-ordinating the action of the various arms, and preparing the detailed operation orders necessary for the direction of a modern army, can view with equanimity the prospect of having to do that work under constant interruption by bombing. And it goes without saying that if we can definitely destroy the head-quarters of an enemy army at a critical moment, we have gone a long way to ensuring the complete paralysis of that army as a fighting body. Moreover, if the head-quarters offices are the brain of an army, its nerves are the _cable communications_ which link the main head-quarters to those of subordinate formations and units, and through which pass the stream of information and orders which enable the body of the army to perform its functions. Not the least important contribution of the Royal Air Force to Allenby's victory at Megiddo in September 1918 was that it

> 'made the enemy's command deaf and dumb—by decisively bombing their main telegraph and telephone exchange at El Afule, a stroke by which Ross-Smith, who later made history by his flight to Australia, helped England to make history. In addition the enemy's two army head-quarters at Nablus and Tul Keram were bombed, and at the second, the more vital, the wires were so effectively destroyed that it was cut off throughout the day both from Nazareth and from its divisions in the coastal sector.'[1]

One result of this, amazing though it may seem, was that the enemy Commander-in-Chief, Liman von Sanders, knew nothing of the collapse of his front or the complete disintegration of the Turkish Seventh and Eighth Armies until the arrival of one of Chauvel's cavalry brigades at his head-quarters at Nazareth, twenty-four hours after zero! A modern head-quarters of any importance is a very formidable affair; for instance G.H.Q. 1st Echelon of our small expeditionary force goes to war with about 800 officers and other ranks, and over 70 motor vehicles; a Corps head-quarters has nearly 400 personnel and over 40 vehicles.[2] An organization of this size cannot

---

[1] Liddell Hart, _A History of the World War_, p. 557.
[2] These figures are approximate only. They include Signals, but exclude motor-cycles.

be hidden in the depth of woods or buried underground, but must be housed in buildings or tents; nor can it be dispersed beyond a certain limit—in fact it constitutes an almost ideal target from the airman's point of view. And although the danger to cable communications may to some extent be minimized by the substitution of wireless, this has very definite limitations; even if wireless sets can be made to withstand the concussion of heavy bombs in their close proximity, wireless communication is still a very long way off being an effective substitute for cable, and if overdone the cure may almost be worse than the disease.

No review of air action against fighting troops would be complete without reference to a form of tactics which in certain circumstances is the most deadly method of attack on this class of objective, namely *Low-flying Action*. It may be defined as attack at point-blank range by light aircraft using small bombs and machine-guns, and taking advantage of high speed, manœuvrability, and natural cover. Certain foreign air forces, notably the American and the Russian, include units specially trained and equipped for this class of work. In the British Service on the other hand we do not favour the special class of 'Assault' or 'Battle' aircraft, mainly on the grounds that it is uneconomical. We regard low-flying action as a special tactical use in an emergency of aircraft which can be, and normally are, otherwise employed, and it is in this sense that the term 'Assault Aircraft' is used in this book. Attempts were made in the War to armour aircraft for assault action, and the Salamander type was a case in point, but these attempts met with little success; nor does it appear reasonably probable that in light aircraft effective bullet-proof protection for the vitals of the machine will ever be practical, except at a prohibitive sacrifice of performance and offensive power. Assault aircraft must rely for protection on surprise, on high speed and manœuvrability— the capacity to jink like a snipe when under fire, and to take advantage of trees, houses, and the folds of the ground. For this reason the single-seater fighter was, and probably still is, the best type for low-flying action; although the light two-seater has one great advantage in that it has a sting in its tail—it can cover its retreat and keep the enemy's heads down by fire during the most dangerous period, the 'pull-up' after releasing the bombs or firing the burst from the forward guns. And it is the fighters, whether single- or two-seaters, that will usually be allotted to this role; though in very exceptional emergencies, such as to turn an enemy's retreat into a rout, all classes other than heavy multi-engined bombers may be turned on to low-flying action, as they were in the instances quoted below. Assault action, however, should not be confused with Dive Bombing. The latter is a special technique, wherein the bomber makes his attack coming down from a considerable height at a very high speed, aiming the aeroplane at the target, and releasing the bomb at a relatively low altitude—from 500 to 1,500 feet; it is a method which is likely to become increasingly popular for engaging strongly defended objectives where the presence of anti-aircraft artillery makes accurate precision bombing from a height a matter of increasing difficulty and danger; but it is a method quite distinct from assault action, which is definitely 'in-fighting'—the 'open sights' of air warfare—in which the pilot carries through his whole attack within a few feet of the ground as long as he is anywhere in the vicinity of his objective.

So the class most commonly used for assault action will be the fighter. But the fighter is primarily an aircraft-destroyer, and his place in normal circumstances is in the upper air, in the front line of the fight for that favourable air situation which alone makes effective action against ground objectives possible. The fighter engaged in assault action is at a serious disadvantage *vis-à-vis* his enemies in the air, for he has sacrificed the primary tactical advantage of height. History has shown, and will prove again, that fighters can only be used against objectives on the ground when a very high degree of air superiority has been firmly established, and when very large numbers are available. At Megiddo we had absolute command of the air, and not a single hostile aeroplane was able to leave the ground—largely because we were so superior in numbers that we could afford to impose a close blockade on the enemy aerodromes. At the Cambrai battle in 1917 we enjoyed a numerical superiority of something like 10 to 1 at the decisive point, on August 8th about 3 to 1. And the fighters of which we disposed on the western front alone in the last months of the War exceeded by nearly 200 per cent. the total numbers in the Royal Air Force of 1934—including the Fleet Air Arm. Again, in the past an essential condition of the employment of assault action on the scale and of the intensity of 1918 was an ample reserve of personnel and aircraft to replace the very high rate of wastage involved. Against troops of the quality and training of the German armies of 1917 and 1918, the casualties involved were indeed on a scale which no air force in the early stages of a war could possibly sustain. For instance, in our attack at Cambrai in November 1917 the casualties averaged about 30 per cent. a day in the squadrons engaged in assault action; and in the March retreat and the Amiens offensive of August 1918 they can have been little less. The astonishing thing is, looking back on it, that any unit of any arm should have been able to maintain its morale and courage in the face of losses such as, for instance, those suffered by No. 80 Squadron in 1918. This squadron, of the 9th (G.H.Q. Reserve) Brigade R.A.F., was employed on assault action almost continuously from the beginning of the March retreat till the end of the War; their average strength was 22 officers, and in the last 10 months of the War no less than 168 officers were struck off the strength from all causes—an average of about 75 per cent. per month, of whom little less than half were killed. With the facilities for replacement of wastage, both human and material, that prevailed after 3½ years of war, a casualty rate on this scale could be met; it would be very different in the opening months of a campaign. It is true that the enormously increased performance of the modern fighter will decrease the rate of casualties likely to be suffered, even against the greatly improved anti-aircraft training and technique that must be common to all first-class armies. Nevertheless, especially in the early stages of a campaign when replacement of personnel and aircraft will be difficult, assault action must be used very sparingly, and retained for conditions of real emergency.

Subject to these conditions there are, broadly speaking, three sets of circumstances when assault action may be justified and will certainly be very valuable. These are *in attack*, to assist the army to break the crust of very highly organized defences; *in pursuit*, to turn an enemy's retreat into a rout; and *in defence*, to hold up the advance of a victorious enemy, and enable our own rearguards to get clear and reorganize the defence.

In the Attack—it has already been stated as a general rule that aircraft are not normally battle-field weapons, but should be used in the enemy's back areas against the movement of his reserves or supply by road and rail. There may, however, be one exception to this rule. It is obviously no use stopping the enemy moving his reserves if his front remains unbroken. The last War proved at a terrible cost the supremacy of the defence on the ground against the traditional arms, and no developments since the War have as yet for certain broken down that supremacy. The tank as a weapon of penetration will probably be the answer, though even against the tank, modern field defences and anti-tank weapons will at least often constitute a very formidable barrier. If land operations on a large scale ever do take place again it is to be hoped that, at least in the initial stages, there will be opportunity for true battle strategy— for manœuvre instead of sieges, for envelopment instead of only penetration. Nevertheless there may be occasions when armies have to attack frontally and break into highly organized defences, such as the great permanent fortifications which look like becoming a feature of all European frontiers, or the strong field defences which, when temporary exhaustion calls a halt to mobile operations, may once more spread themselves across the face of great areas of country. If these conditions do arise, then it may again be necessary—as it was on August 8th, 1918—to concentrate every available effort, including that of the assault aircraft, temporarily on ensuring the initial break-in without which the subsequent exploitation is obviously impossible. In other words, in order to make sure of *breaking the crust* of a highly organized defensive system we may again have to employ our assault aircraft *temporarily* in close support of the armoured force that will be used to break in on the ground, against such objectives as artillery areas, anti-tank weapons, defended road blocks, or the movement of the enemy's immediate reserves on the battle-field. *But this must only be temporary*; and as soon as the break-in is made and the assaulting troops on the ground have penetrated the enemy's defences, then the assault aircraft must at once be lifted against the more suitable and important objectives well in the enemy's back areas, with the aim of paralysing the movement of his main reserves and converting the break-in into a real break-through.

It is, however, in the *Pursuit* against an already broken enemy that the supreme opportunity for assault action will arise—so much so that a new and more complete value seems to attach to the old and much abused principle of annihilation of the enemy's army as the crowning achievement in battle. Here if ever will be the occasion on which the victorious commander will be justified in accepting all risks, and diverting all his aircraft of each and every class to set the seal on his success, and convert his enemy's retirement into a rout. On three different fronts the last months of the Great War turned a new page in military history. In Palestine, the story of the annihilation of the Turkish Seventh and Eighth Armies in the pursuit from Megiddo is well known, but T. E. Lawrence's description deserves quotation:[1]

'But the climax of air attack, and the holocaust of the miserable Turks, fell in the valley by which Esdraelon drained to the Jordan by Beisan. The modern motor road, the only way of escape for the Turkish divisions, was scalloped between cliff and precipice in a murderous defile. For four hours our aeroplanes replaced one another in series above the doomed columns: nine tons of small bombs or grenades and fifty thousand rounds of S.A.A. were rained upon them. When the smoke had cleared it was seen that the organization of the enemy had melted away. They were a dispersed horde of trembling individuals, hiding for their lives in every fold of the vast hills. Nor did their commanders ever rally them again. When our cavalry entered the silent valley next day they could count ninety guns, fifty lorries, nearly a thousand carts abandoned with all their belongings. The R.A.F. lost four killed. The Turks lost a corps.'

Actually on the same day (Sept. 21st, 1918), on the Macedonian front, a similar fate befell the Second Bulgarian Army.

'In the Kosturino Pass the retreating enemy was scattered time after time, and men, transport, and animals blown to bits. Wagons were lifted off the road and flung down ravines. But the greatest execution of all was done in the narrow Kresna Pass,[2] a wonderful defile through which the Struma river comes down

[1] See *Revolt in the Desert*, p. 392.
[2] This actually was almost an ideal bit of ground for the purpose—not unlike the valley of Glenshiel in the highlands of Scotland.

from Sofia, and up which the Struma army was escaping. Here as elsewhere our aviators flew as low as 20 feet above the fugitives, machine-gunning constantly and killing hundreds. This target was sixty miles from the most advanced aerodrome, and with mountains of over five thousand feet between.'[1]

'From all sides fires broke out, guns were abandoned in gullies, rifles, equipment, baggage were thrown away, and the demoralized army fled towards its homeland.'[2]

Finally, on the Italian front, the sequel to Vittorio Veneto was another and equally ghastly example of the terrible potentialities of air pursuit.

'It is doubtful whether in any theatre during the Great War the R.A.F. obtained such targets as those which were offered to them in Italy on the 29th and 30th of October. It is certain that they took full advantage of the opportunities presented to them. On these two days the Conegliano–Pordonone road was black with columns of all arms hurrying eastward. On to these the few British squadrons poured thirty thousand rounds of S.A.A. and three and a half tons of bombs from low altitudes. Subsequent examination of the road almost forced the observer to the conclusion that this form of warfare should be forbidden in the future.'[3]

It has been pointed out by Sir F. Maurice[4] that Clausewitz and Henderson both classed the decisive pursuit as a principle of war, and that owing to the increased strain of modern war and the delaying power of modern weapons there had been, until the closing months of the last War, no decisive pursuit since 1870. The same writer has suggested that the air pursuits described above indicate that the air arm has restored the pursuit to its former dignity as a principle of war. Whether or not much importance need be attached to these codified principles, there is no doubt that decisive pursuit, to the extent of the virtual annihilation of a defeated army as a fighting organization, has definitely taken its place once more as a practical operation of war. But it is necessary to add a word of warning. The Field Service Regulations[5] stress the importance and value of cavalry

[1] Collinson Owen, *Salonica and After*, p. 261.

[2] Luigi Villari, *The Macedonian Campaign*, p. 233. See also Seligman, *The Salonica Sideshow*, pp. 135–6, and Sir George Milne's dispatch dated Dec. 1st, 1918.

[3] Maj.-Gen. the Hon. J. F. Gathorne-Hardy, *Army Quarterly*, October 1921.

[4] See *British Strategy*, p. 32.   [5] F.S.R., sect. 76.

in the pursuit. Now cavalry are supremely vulnerable to assault action from the air; and it is quite certain that in the conditions envisaged the enemy aircraft will be exerting themselves to the full to delay the pursuit—except when an almost complete victory in the air has accompanied success on the ground, which we cannot assume will necessarily be the case. If, therefore, we are to launch these very vulnerable horsed units in the pursuit, there will inevitably be an almost irresistible call for their protection against hostile assault action, and this will very likely mean that the fighters—probably the most effective arm in the pursuit—will be diverted from the pursuit to protect the cavalry, at a time when every aeroplane in the air should be employed in harassing the retreating enemy, or at least affording protection to the aircraft engaged on that duty. The moral surely is that once more the armoured man in a fighting vehicle must replace the unprotected man on a horse; that the pursuit must be carried through from the air by the assault aircraft, and on the ground by the tanks, who are the arm least vulnerable to air action; and that the main bodies of the pursuing army must temporarily accept the disadvantages and dangers of such hostile air action as can be brought against them, minimizing those effects by concealment and dispersion by day, and movement mainly by night.

The third occasion for the use of assault action, namely to deal with a situation of real emergency in *Defence*, is one which may be of diminishing value as the methods of attack on the ground adapt themselves to modern conditions. Moreover, it is an occasion on which the decision to use aircraft in this way requires the very nicest judgement, since we must anticipate that the enemy will be endeavouring to exploit his initial success by the methods described above; and the protection of our own forces on the ground may temporarily become of the first order of importance. Nevertheless there may still be occasions, such as the situation at Le Cateau in 1914 or the German offensive of March 1918, when the one overriding consideration, to which all others must take second place and for which all risks must be accepted, is to stop a hostile break-in, to 'putty up' a gap until reserves can arrive on the ground, or to enable hard-pressed rearguards to disengage and reorganize. There were several occasions of this sort in the last War, when the value of low-flying action was amply proved; the first being the German counter-attack at Cambrai in December 1917, when for the first time our fighters came into action against advancing German infantry, to cover the withdrawal of our troops from forward positions which had become untenable. It was, however, during the retreat of Gough's Fifth Army in the following March that there took place the most bitter and intensive rearguard actions ever fought by British aircraft, with results the importance of which it is difficult to overstate, and at a terrible cost in casualties. As a matter of fact even at that time there were complaints that the use of the fighters in assault action left our army co-operation aircraft at the mercy of the enemy fighters, and so deprived the artillery of the observation that is so essential in these conditions. Actually there is not much substance in this criticism, and the absence of air co-operation with the artillery was, as often as not, due to failure to get the battery wireless sets into action;[1] and it is only quoted in order to indicate one of the incidental disadvantages that may attend this use of fighters. There is, however, no question whatever but that the action was amply justified by the results —especially on the 25th and 26th of March, when a break-through west of Bapaume was averted, and a gap between the French and British about Roye was stopped by a concentration of every available fighter on low-flying action, which (in the words of a senior General Staff officer at the time) 'froze up' the German advance on each occasion. These operations are described in full in the Official History[2] and need no further elaboration here. But it is worth noting the extent to which the value of assault action in defence was impressed upon the minds of the High Command at this critical time. In the directive issued by General Foch to the Allied forces in France on April 1st, 1918,[3] it was actually laid down that 'At the present time, the first duty of fighting aeroplanes is to assist the troops on the ground, by incessant attacks with bombs and machine-guns, on columns, concentrations or bivouacs. Air fighting is not to be sought except so far as necessary for the fulfilment of

---

[1] See *Military Operations, France and Belgium, 1918*, p. 168.
[2] See *The War in the Air*, vol. iv, pp. 323–7; also p. 380 et seq. for the low-flying action at the Lys battle in April.
[3] Appendix XVIII to *The War in the Air*, vol. iv.

this duty.' This is putting it very strongly—far more so than could ever be justified in circumstances of any but the gravest emergency. The reader must note the qualifying phrase 'at the present time', and remember not only the desperate nature of the situation on the ground at that time, but also the situation in the way of relative strength and facilities for replacement, already mentioned on a previous page.

Whether the use of low-flying tactics in defence will ever again have an influence comparable to that which it exerted in 1918, when the objectives were mainly masses of infantry, their transport and supporting artillery, is perhaps doubtful. If it could be assumed that an infantry battle on the 1915–18 lines will ever be fought again, then again the low-flying fighter would undoubtedly be brought in in the last resort to assist the rearguards on the ground. But a clearer estimate of the probabilities can be gained by considering the possible effect of low-flying action against our tank break-in at Cambrai, or on August 8th; and it is difficult to convince oneself that this effect could have been very serious. And if, as seems more probable, the role of infantry in a future war tends increasingly to become that of a holding and consolidating force, while the break-in and subsequent exploitation is carried through entirely by the armoured units, then there seems little doubt that the extent to which assault aircraft can help the defenders will be correspondingly reduced.

Finally, there emerge from the experience of the last War certain points on the subject of *the control of assault squadrons*, which seem certain to arise again in any future campaign. The first is that pilots engaged on this duty must not merely be sent out with a roving commission to shoot up what they think fit, but must be given clear and definite orders before going into action. This is not to say that they should be tied down absolutely rigidly and given no discretion at all—obviously they must be allowed to use their initiative and common sense, and like any one else in action, in the air or on the ground, must even be encouraged to depart from the letter of their orders if the situation obviously so requires. But the military commander ordering the assault action alone can know the effect he desires it to produce; and he should not only make that clear to the units concerned, but must also lay down quite definitely the types of objective to be engaged, in an order of priority, and must delimit the areas within which different units are to work. In this he will have the assistance and advice of the air force commander, who in his turn is of course entirely responsible for the method by which these orders are carried out—such details, for instance, as the strength of formations to be used, and the number and type of bombs. This, perhaps, is all rather obvious, but the point is stressed because in the early days of assault action, during the summer battles of 1917, the orders were not nearly definite enough. In the attack at Messines in June, for instance, the low-flying pilots of the 9th Wing 'were to rove at their will and shoot at any troops, guns, or transport which they discovered'.[1] Again, the orders for the 5th Brigade for the third battle of Ypres laid down merely that 'a vigorous offensive will be carried out by machines of both wings (i.e. 15th and 22nd) against all favourable enemy targets on the ground West of the line Staden–Dadizeele'. These orders were very gallantly carried out, and the Official History contains the description of several 'crowded hours of glorious life' which ensued. But it cannot be claimed that in either battle these gallant and costly operations really had any considerable influence on the issue. The Air Force command were not slow to learn from experience, and the 3rd Brigade orders for the attack at Cambrai four months later were much more definite and detailed, both as to the type of target to be attacked, the timing of the attacks, and the allotment of areas to units.[2] Instructions on similar lines were issued to the 22nd Wing for the attack on August 8th, 1918, with results that are described in a later chapter; and in both battles the operations of the assault aircraft had a really important influence on the result. The Germans had specially trained units for this duty, known as 'Battle Flights', which they used on occasion very effectively, notably in their counter-attack at Cambrai on November 30th, 1917.[3] And in a General Staff pamphlet issued in February 1918 on the subject of the employment of these battle flights[4] they laid particular stress on the necessity for clear and definite orders.

---

[1] *The War in the Air*, vol. iv, p. 129.
[2] See Appendix XI to *The War in the Air*, vol. iv.
[3] See *The War in the Air*, vol. iv, p. 251.
[4] See Appendix XII to *The War in the Air*, vol. iv.

The second point is that these clear initial orders are not enough. They are essential for the aircraft that go over at zero in support of the attacking troops on the ground; but as the battle continues and the situation develops, fresh aircraft will be thrown into the fight, and they, too, will require equally definite orders. These are obviously much more difficult to produce, mainly because after the initial attack the situation is nearly always confused, and often quite unknown to the commander concerned. The difficulty is enhanced by the fact that the most favourable opportunities for assault action will often be fleeting, by the rapidity with which assault aircraft are committed, and by the virtual impossibility of altering or modifying their orders once they have taken off. No doubt after the first assault subordinate commanders, squadron leaders and flight commanders, will have to be given a freer hand, based on a clear understanding of the effect to be produced; but it will still be necessary as far as possible to direct the low-flying attacks against definite known objectives. And this is obviously a question first of getting the necessary information quickly, and secondly of a good system of intercommunication by which that information and the subsequent orders can immediately be conveyed to the squadrons. When the objectives are the most suitable ones, namely the movement of reserves in the back areas, then the information will all have to come from the air, probably from specially detailed reconnaissance aircraft. But in the exceptional circumstances when assault aircraft are operating in close support of attacking troops on the ground, the problem of quick and accurate information is more difficult; and it may be advisable on such occasions to place the assault squadrons temporarily under the orders of relatively subordinate army formations, such as corps, which are likely to be the highest formation in a position to obtain the necessary information sufficiently rapidly. In these circumstances the commander will depend to some extent on the reports of reconnaissance aircraft; but these will seldom be enough in themselves, and cannot be an adequate substitute for reports from the forward troops, especially in close country. It is to be hoped that modern developments in wireless intercommunication on the ground will provide the solution to this problem. Without some such solution assault action in close support will always be severely handicapped. Pilots in the heat of action, flying at tremendous speed a few feet above the ground, cannot be expected to appreciate the niceties of the tactical situation on the ground; and in the absence of information and definite orders will often expend their energies, and even their lives, attacking targets of minor importance, while the objective which is really vital at the time remains untouched. One such instance was the hold-up at Flesquières on November 20th, 1917, the first day of the Cambrai battle, which is dealt with in some detail in the Official History.[1] Throughout this day the village of Flesquières held out, though our troops had penetrated deeply on either flank; and a large number of British tanks were destroyed by direct fire at short range from German batteries closely sited just behind the Flesquières ridge.[2] This artillery area was among the objectives detailed to be dealt with by assault aircraft at zero; but after the initial attack it received no further attention, and the assault pilots 'spread their attacks over German troops, transport, and other targets of a general nature'—no doubt for the good reason that no one, least of all the fighter wing commander, had any idea of the real situation at Flesquières. It is a little strange to find the Official History saying 'even had the fighting pilots known of it and realized its importance, it would be idle to claim that their attacks could have been made powerful enough to wipe out the German resistance'; and going on to suggest that artillery in the open is not a particularly suitable target for assault action. With this the present writer takes leave to disagree. On the contrary, batteries in exposed positions are particularly vulnerable to assault action. It is certainly difficult to get a direct hit on a gun with a bomb, or to do it much damage with a machine-gun; but the detachments are completely exposed to attack from above and behind. And if the available assault aircraft on this occasion could have been concentrated on this group of batteries, there appears little reason to doubt that their resistance could have been overcome, with tactical results of the utmost importance. Be that as it may, the

---

[1] See *The War in the Air*, vol. iv, pp. 234–8.

[2] Unfortunately the romantic story of the single German gunner officer serving his last gun, which was given such publicity by our own communiqué at the time, has since joined the 'corpse-factories' and other myths of the Great War.

fact remains that they were *not* so concentrated, because the information about the situation did not get back in time; and the incident serves to illustrate the importance of information in these circumstances. This factor again is given prominence in the German Staff pamphlet already referred to, and has practical corroboration in the report rendered by the officer commanding the 22nd Wing R.A.F. after the battle of Amiens in 1918:

'It is essential for all pilots so employed to know their area well, and have a thorough insight into the situation. A time comes when the enemy will reorganize his defence and our progress will slow down. This is a time when low-flying fighters must be sparingly employed, and then only on reliable information on the situation at points where local counter-attacks are developing, or where our troops are held up at strong points.'

The last point is one which is referred to in the report just quoted, namely the importance of low-flying pilots knowing the ground over which they have to operate. This again is a condition which is principally important in the exceptional circumstances when assault aircraft are acting in close support of troops on the ground—though it is always desirable even when the objectives are such as columns on the roads well in rear of the actual battle area. It is, to say the least of it, taxing pilots very high to expect them to fight practically at ground level, and at enormous speed, over country which they have never seen before. But when they are required to operate actually on the battle-field, in close support of the other arms, where a few hundred yards or a few seconds may just make all the difference, it is not too much to claim that a knowledge of the ground is essential. And this again is recognized in the German Staff memorandum on the employment of battle flights, which goes so far as to say that 'accurate knowledge of the ground is the first condition for the successful action' of this class of unit.

SUPPLY

So much for a necessarily short and incomplete outline of the subject of air action against fighting troops; that of the dislocation of *Supply* can only be dealt with in still more general terms. It is obviously impossible to attempt anything like a

detailed survey of the whole administrative system o
from the point of view of its vulnerability to air actio
be sufficient to sketch in very roughly the outlines of
bearing in mind that the details, the actual objectives ιυ
attacked, will vary in every case and will always involve that
intimate knowledge of the enemy's administrative machine, the
necessity for which has already been referred to in this chapter.

The measures adopted to dislocate an enemy's system of
supply will not differ very much from those directed against
his fighting troops, and amount in the main to action against
communications—rail, road, and water. So that during battle
periods on the ground the operations of the air striking force,
designed primarily to interfere with the movement of hostile
reserves, will at the same time have an important effect on his
service of maintenance. And therefore—other things being
equal—the air plan of battle should provide for the maximum
possible dislocation of the enemy's system of supply, consistent
with the achievement of what must be the primary object at the
time, namely the paralysis of enemy troop movement. Probably
more often than not the two types of objective will be in the same
area, using the same channels of communication, and action
against the one will automatically affect the other.

But a more interesting range of possibilities is opened up by
reflection as to the possible results of air action against supply
during periods other than when an actual battle is in progress
on the ground—in fact during the long stages of deadlock and
preparation for fresh activity which are bound to occur in any
land campaign of any importance. These 'periods of comparative inactivity' may not take the same form as they did in the last
War, when huge armies lay facing each other for months on
end across a narrow strip of no-man's-land, on fronts measuring
hundreds of miles in extent. But unless the war is over in a
matter almost of weeks—and very, very few wars have ever been
over in a matter of weeks—there are bound to be such periods,
no matter what may be the nature of the forces engaged.
Armoured and mechanized forces cannot be continually fighting,
any more than could the old 'foot and horse' armies, indeed
probably less so. Whether the system of maintenance of such
modernized armies will be on the same lines as that of the
traditional army is open to question; and that their supply

requirements in terms of daily tonnage, especially of ammunition, will be much smaller appears almost certain. Most first-class armies, notably that of Russia, are to a greater or less degree increasing the size and efficiency of their tank corps and substituting the motor vehicle for the horse; but it is true to-day, and probably will be for some time to come, to say that all armies are organized and equipped mainly on the traditional model, with infantry as the backbone and bulk of the army, supported by a powerful artillery. It is still the fashion to measure the strength of armies in man-power; and if there is one lesson above all others which emerges from the last War, it is that man-power can only be effective if it is supported by overwhelming shell-power. The presence in modern armies even of their comparatively small tank forces should reduce the proportion of shell-power required to enable them to advance—it is to be hoped that we shall not again need to expend $5\frac{1}{2}$ tons of high explosive for every yard of front attacked, to enable us to capture a single ridge on a front of a few miles, as we did at Messines in 1917. But, as long as armies remain on the man-power and shell-power basis, their daily requirements in food, ammunition, forage, petrol, engineer stores, and material of all kinds will still be measured in thousands of tons and dozens of train-loads.

The question of what influence might be exerted by air attack against the administrative services of an army of this nature was the subject of an interesting article[1] in the *Army Quarterly* for January 1926 by Colonel Wingfield, who was a Q Staff officer at the Head-quarters of the Second Army in the latter stages of the War. In that article it was pointed out that economy in transportation required the delivery in bulk of each commodity, such as meat or forage, to base depots in the theatre of war, where working reserves were held to compensate for the irregularity of supply. These base depots were sited in the vicinity of a line dividing the rear, or producing and collecting, zone from the forward, or distributing, zone—whether that line was, as on our own side, the Channel coast, or an arbitrary line fixed by the Administrative Staff as it was in continental armies. 'The point is that reserves of considerable size in bases are absolutely necessary for the working of the administrative

---

[1] See 'Air Operations against the Lines of Communication of an Army', by Colonel the Hon. M. A. Wingfield, *Army Quarterly*, Jan. 1926.

machine.' Colonel Wingfield concluded that the danger-spot in an army's administrative system lies in front of its base depots, where there are held only small emergency reserves; he drew this conclusion from a description of the administrative situation of our own Second Army in Flanders in the spring of 1918, based no doubt on an examination of that situation which was held at Army Head-quarters at the time. The situation revealed by that investigation was indeed of great interest. The Second Army at the time comprised twenty divisions, in the line and in reserve, east of a line drawn north and south through St. Omer.[1] The base depots were at Calais and Boulogne, and the army front was served by two double lines through Bourbourg and St. Omer and one single line through Blendecques; the capacity of these lines being forty trains a day for each double line and eighteen for the single line, a total capacity of ninety-eight trains per day between the Second Army railheads and its bases. The total requirements during active operations were estimated by the Second Army Staff at sixty-eight trains a day for all services, not including civil traffic, railway maintenance, or abnormal troop movement, representing nearly three-quarters of the total capacity of the lines. Forward of the base depots there were in the army area at any one time a total of five days' stocks of supplies and ammunition, in the field depots shown on the map and with the troops themselves. There can be no doubt that, when the aim is to interrupt supply to the forward troops during active operations, the place to do it is not behind the base depots, thus leaving available reserves amounting, in the case of the British army in France, to between thirty and forty days' consumption. What is not always so obvious is where in front of those base depots the blocks should be made, and the correct decision on this point will again depend on an accurate knowledge of the enemy's supply system, including the location of his intermediate depots, the condition and quantity of his motor transport echelons and of the roads leading to his forward areas. As Colonel Wingfield points out, the obvious places in the Second Army area appear at first sight to be at Hesdigneul and the junction south of Calais; but in fact blocks at these points would still have left five days' stocks available in front of them, whereas blocks at Bourbourg, St. Omer,

[1] See Sketch-map 1.

and Arques would have completely cut off the divisions in the Army area even from the three days' reserves in the field depots. The conclusion arrived at was that if blocks at these points could be maintained for six days the situation would have been critical; it seems in fact that less than half that period could have been more than critical. As a matter of fact it is more than doubtful if the twenty divisions of the army could have been supplied even if only one of the double lines had been put out of action, and they certainly could not have subsisted on only one line, if only because of the difficulty of distribution from railheads. And it is uncomfortable to contemplate the situation which might have arisen if the Germans, in conjunction with their April offensive on the Lys which got them nearly to Hazebrouck, had been able to put down a really intensive and sustained air attack on these railways, which even as it was were strained to their utmost capacity to deal with the reinforcing divisions which were rushed in to stem the enemy's advance on the Channel ports.

But there is another factor which has to be considered in selecting objectives of this sort. While it is desirable if possible to make the blocks on the railways sufficiently far forward to cut off the forward troops even from their field depots, they must not be made so far forward as to enable the enemy merely to substitute road for rail transport in front of them. The Second Army area east of the line Bourbourg–St. Omer was not well served by roads, and actually the plan was to flood it in the event of a serious German attack; and there was no doubt that neither the roads nor the motor transport at the disposal of the army would have stood the strain indefinitely of drawing everything required from west of St. Omer. The road transport problem is always a serious one during active operations, and to throw back railheads even a comparatively short distance would have serious results. Actually, blocks at the three places mentioned would have thrown back railheads about twenty miles, and the effect would very likely have been fatal. The capacity and elasticity of modern roads and motor transport no doubt is greater than in 1918; but the principle remains the same, and is that to interfere effectively with supply the railways must be cut sufficiently far back to impose an intolerable strain upon the road transport echelons.

This study of the dangerous situation on the lines of communication serving our own Second Army front suggested an
examination, on the same lines, of the position in the back areas
of the German army holding the same sector. This revealed
four vulnerable points on the enemy's railway system opposite
the Second Army—two at short range, Armentières station and
the railhead between Comines and Wervicq, and two at a greater
distance, Tournai station and the railway works south-west
of Courtrai, the destruction of which would have the effect of
throwing practically all the enemy's broad-gauge traffic on to
the single line between Avelghem and Roubaix. In spite of
the comparative inefficiency of air bombardment in those early
days, General Plumer, the Commander of the Second Army,
had great confidence in its capacity for damage and dislocation,
based presumably on his own experience of the relatively slight
German efforts against our own back areas. He was convinced
that serious dislocation could be caused to the enemy's administrative system, both during active operations and during the
periods of preparation for active operations, by the bombing of
carefully selected objectives on the enemy's railway system. But
he was equally emphatic that an essential condition of success
was the concentration of all available aircraft continuously on
the minimum number of points where the greatest dislocation
could be produced, targets of lesser importance being temporarily neglected. Unfortunately his advice, and the instructions
of General Foch on similar lines quoted in a later chapter,[1]
were not followed—no doubt for reasons which appeared sufficient at the time, but with what results in the shape of lost
opportunities we can now only guess. It is significant that it was
on General Plumer's front that the Germans in an official report
ascribed the failure of their spring offensive in 1918 partly to
'the difficulty of supply under the increasing attacks from the
air'.

The claim that 'the maintenance of an army in the field may
be prejudiced—even totally prevented—if the enemy's aircraft
is successfully handled' is one that has been seriously stated,
not by any unbalanced air enthusiast, but by an administrative
staff officer of great experience who has been faced in war with
the vast complications of the maintenance of a great army.[2]

[1] See p. 128 below.   [2] See Colonel Wingfield's article, p. 378.

Whether that claim can be substantiated time alone will show, though no one could contemplate with any degree of comfort the prospect of having to supply an army of the traditional model in the face of modern air action. Fortunately it is not within the province of the present writer to offer a solution of that problem, should it ever have to be faced. But whether the force to be maintained is one on the old man-power and shell-power basis, or whether it is a new model army of armoured and motorized formations, it does seem that the threat from the air calls for a new technique of supply, the details of which can be worked out only by experienced experts and administrators as the result of trial and experiment. It must be sufficient here to indicate some few of the lines on which such experiment might be directed, and suggest some of the features of our present administrative system which may have to be modified. It may be worth anticipating an almost certain criticism by admitting at once that some of the modifications suggested in this system will be difficult, uneconomical, and inconvenient. Of course they will. The present system was evolved before the threat from the air became what it is to-day, and took its present form because it was found that that form was the easiest, most economical, and most convenient—*in conditions which are now out of date.* We must be prepared to change our system where necessary to meet changed conditions; and if we try to adhere to our proved methods simply because they worked well in the last War, we shall have a rude awakening in a future war to find our whole system breaking down.

Starting with the most distant link in the chain of supply: the first feature of the system which suggests itself as especially vulnerable is the bulk delivery of each commodity from the producing zone to the bases in the theatre of war. The loss of even one shipload of some important commodity may have a very widespread and damaging effect at the front. We can avoid shipping in bulk certain more specialized stores, such as aircraft or engine spares and certain classes of ordnance stores; but in the main this is a risk which has to be accepted. In most cases it is obviously impracticable to break bulk in the collecting zone. The bulk shipment of most commodities such as ammunition, forage, meat, petrol, even motor transport, is essential if only because many of them require special classes of ship; and we must hope that the navy and the air forces co-operating with the navy will be able to afford adequate protection on the voyage. Again, a base port will constitute a very important and vulnerable objective: appalling dislocation and damage could be inflicted not only by bombing the port itself, sinking ships in the fairway or alongside the quays, but possibly almost as much by disorganizing the facilities for rail clearance from the dock area. We cannot hope always to have our base ports out of hostile air range, indeed as years go on we shall probably never be able to do so. But we should face the fact that we shall never again be able to site our base ports as close to the front[1] as, for instance, were Calais and Boulogne behind the Second Army area; and that wherever they are (except on the very rare occasions when they can be out of range) they will make a serious demand on our resources for anti-aircraft protection.

There appear to be three measures—two of them unfortunately rather conflicting—which we can adopt to minimize the risks of bulk delivery and storage in base depots. The first, which follows logically on what has been said about base ports, is that *our base depots must be much farther from the front* than they were in the last War—and, indeed, are frequently envisaged in peace exercises to-day. This, of course, is a serious nuisance; it means waste of time, large additional demands on the already strained resources in rolling stock and locomotives, additional troops for guarding vulnerable points on railways, and probably additional intermediate depots forward, holding reserve stocks to provide against possible interference with the railway. But it will have to be accepted. The lay-out of base depots can be dispersed up to a reasonable limit, but we cannot possibly afford to have all our reserves of supply and ammunition within close and easy bombing range as we did in the last War. Secondly, it may be necessary to hold *larger reserves in the theatre of war*; the number of days' stocks that it may be necessary to hold must be decided in every situation on its merits, according to a number of factors including the length of the sea voyage involved and the scale of naval and air attack to which shipping may be

[1] As long ago as 1915 the primitive British aircraft of those days forced the Turks virtually to abandon their base port of Ak Bashi Liman; and the early successes of the old Short torpedo seaplanes against Turkish shipping in the Dardanelles completed the work of the submarines in cutting the enemy line of sea communication to the Peninsula. See *The War in the Air*, vol. ii, pp. 64–72.

liable. Neither the navy nor the air force can ever guarantee immunity for cargo ships at sea—indeed, the one measure which gave real security against undersea attack, namely convoy, is a definite disadvantage and provides a target for air attack far more conspicuous and vulnerable than single vessels. Therefore, since the chances of interference with the regularity of supply are increased, so the reserves held to provide against that contingency should be correspondingly increased.

Lastly, although it seems probable that we may have to increase our reserves in the war zone, it is unfortunately certain that we shall have also to *decrease the size of the depots in which these reserves are held, and increase the number of depots.* In short, instead of having all our larger number of eggs in one larger basket, we shall have to have a lot of smaller baskets. General Plumer, as we have already seen, was for using the bombers to cut railways rather than for destroying dumps of ammunition or other stores; in his view, for instance, the destruction by German bombers of the ammunition depot at Campagne on the night of the 18th–19th May, in which the damage inflicted amounted to only 30 casualties to personnel and 1,000 tons of ammunition destroyed, was of minor importance. It must be remembered that this was in the atmosphere of 1918; and to the man who had accumulated 255 trainloads of ammunition for his very successful but limited attack at Messines the year before, the loss of a mere three trainloads must have appeared to be of much less account than it would appear to the administrative staff in the early stages of a future war! Nevertheless General Plumer was undoubtedly right in saying that actually during a battle the loss of one dump was of far less importance than the results of a corresponding amount of effort directed against forward rail communications. In any circumstances the loss of a single minor dump of ammunition or any other store, though possibly embarrassing, would be little more. It would be quite another matter if it were one episode in a widespread plan of operations designed, during a period of temporary deadlock on the ground, to make the position of an enemy army untenable by the piecemeal destruction of the supply organization behind it. In these circumstances even the smaller reserve depots and dumps in the forward area may be well worth attacking. At Campagne only 1,000 tons of ammunition

went up; at Audruicq about two years earlier four German
bombers destroyed 8,000 tons, together with 23 sheds and about
a mile of railway track. Losses on this scale occurring not as
isolated incidents but as part of a thorough, co-ordinated plan
must soon have become very serious even in the luxurious days
of 1918; while in the early stages of another war, when our
ammunition situation even without them is bound to be far
from easy, they would surely prove fatal. Both Audruicq and
Campagne were only minor depots, and the results of their
destruction were correspondingly unimportant at that time—
though the loss of 8,000 tons of ammunition in the early stages
of another war would be anything but unimportant. The far-
reaching effects of the loss of a really large quantity of ammunition are illustrated by the explosion in the German dump at
Spincourt[1] which destroyed 450,000 heavy shells and, in the
view of the French General Palat, was a very important factor
in the saving of Verdun. So there is no doubt that we can
afford to take no risks, and whatever the inconvenience involved
we must split up our base ammunition depots into more and
smaller units, all separated as far as possible from each other,
and each dispersed over as wide an area of ground as possible
so as to minimize the effects of bombing.

Ammunition, of course, is the most vulnerable form of
supply, with the possible exception of petrol. As and when
heavy oil replaces petrol as a fuel for road vehicles and aircraft,
so will fuel reserves become correspondingly less vulnerable.
The problem of the protection of petrol echelons in the forward
areas has been touched upon already in this chapter. At the
base it is still more difficult, the more so because in all developed
countries petrol is delivered direct from tankers at certain ports
into bulk installations ashore, and the tendency is increasingly
to replace cans and drums—which can be dispersed in depots
inland—by the bulk wagon on the railway or road, drawing
from the bulk installation at the port and delivering direct to
railheads, aerodromes, or forward echelons. These dangers
cannot always be minimized by dispersing the delivery from
tankers to a number of different ports, because ports like ships
are designed to deal with specialized forms of cargo and not
all of them are fitted to receive petrol in bulk. The difficulty

[1] Not due to air action.

might be eased by mooring tankers alongside the quay at several different ports and using them as temporary base depots—though the disadvantages of that are obvious in view of the inevitable shortage of tanker tonnage in war. Or it may be practicable to use great tank barges moored up rivers—though no port authority will welcome the possibility of one of these barges being hit and burning petrol coming down on the stream. In any event petrol stocks in a theatre of war will always be a source of anxiety to the administrative staff and the air defence commander, and it all points to the urgent need for the rapid development of light engines using the less inflammable heavy oil fuel. Frozen meat is another form of bulk supply which it is much more convenient and economical to store concentrated at one base, and here again we should not in future rely upon one single great cold-storage plant commandeered or built at a single port. But although these specialized forms of supply are, perhaps, the most actually vulnerable and easily destroyed, still all other commodities—rations, forage, engineer stores, ordnance stores, aircraft spares, motor-transport reserves, remounts, and so on—must all be dealt with on similar lines and dispersed over more and smaller depots. It is true that they are less easy to destroy, especially if they are properly spaced out on the ground within the depots; still, interference with their supply to the fighting troops can be effected not only by material damage and actual destruction but by dislocation, by disorganization of issue and distribution, by constant dowsing of lights at night and moral effect upon the personnel working the depots.

Now two of these factors in combination—the need for base ports to be farther from the front, and for more and smaller depots—do suggest a possible solution, or at least a line of experiment and development, which might at least increase the difficulty of effective dislocation of supply from the air. For the normal series of large base depots for each class of supply, located in the vicinity of the ports—or of the rear boundary of the distributing zone where the front is far from the sea—and holding the bulk of the reserve stocks, with only small field depots forward on a comparatively short line of communication, it seems not impossible to substitute a number of smaller depots suitably spaced along the longer line of communication of the future. Thus, for instance, instead of big base depots at Calais and Boulogne holding thirty days' supply, with small field depots at Audruicq and Zeneghem holding only two or three days' supply, serving the area east of St. Omer, it would be necessary to have the base port on the Atlantic coast; and along the line of communication there might be five or six smaller depots each holding—say—eight to ten days' supplies; the whole providing an ample reserve against interruption of bulk shipments, and each so sited as to give a reasonable margin of safety against temporary interruption of the rail communications on either side of them. All this, of course, would mean added complications; it would be uneconomical; it would require more space, which is often difficult to find in some theatres of war; it would require more personnel, both for working the depots and for the necessary guards; it would mean an additional strain on the railways, more difficulty in marshalling and making up pack trains, and probably more dead mileage for rolling stock. But if it is safer—if it or something like it is going to enable the administrative machine to continue functioning in the face of air action—then all these difficulties and disadvantages must be faced and overcome.

No doubt the experienced administrator will be able to suggest many other and better ways of meeting the menace to supply from the air. It may be possible to increase the elasticity of the supply organization by a still wider use of motor transport —though the road can never replace the railway, and a heavy use of motor transport beyond a certain point is liable to start a vicious circle, involving as it does huge tonnage requirements in such commodities as roadstone, petrol, and spare parts. We may be able to supplement the broad-gauge railway by the light railway of the decauville type, which is quickly and easily laid; or in some countries by large-scale development of inland water transport. But the writer hesitates to rush in farther, where angels fear to tread. These matters are the business of the expert; and provided the expert realizes the extent and scope of the air threat to supply, and is not content to sit back and admire the present system of maintenance merely because it has worked well in the past—then there are grounds for hope that our national genius for organization will find a solution to these very pressing problems.

VII

AIR ATTACK ON COMMUNICATIONS

It will be apparent from the foregoing chapter that, whether the object at the time is to strike at the fighting troops themselves or to deprive them of their essential supplies, the method in the main amounts to interference with *movement*, by striking at the communications, rail or road, serving the battle area. The aim of the following pages is to describe in a certain amount of detail some of the vulnerable points on a line of communications, to suggest how they may best be attacked, and to suggest tentatively some of the results that may be expected to accrue from such attacks. When considering this subject there are two factors which the reader should bear constantly in mind. The first is one which has already been stressed in a previous chapter, and is the importance of a *proper use of all available intelligence, supplemented by expert technical advice.* The fact that we cannot effectively interfere with an enemy's system of transportation unless we have an accurate and detailed knowledge of that system is sufficiently obvious; and it is hoped that the examples quoted in this chapter—most of which have emerged from consultation with practical, working, transportation officers —will make it equally clear that information alone is inadequate without skilled technical advice as to the vulnerable points and weak links in the system to be attacked. The second point to remember is that *air action in this sphere—as in most others—depends for its effect far more upon dislocation and disorganization than upon actual material damage.* It is perfectly obvious that we can never hope to demolish wholesale great railway junctions or long stretches of road; nor indeed is it necessary. It is axiomatic that in any system of transportation, but of course especially on railways, there must be smooth working and a rapid turn-round—a balanced movement of vehicles in each direction. This is not merely a matter of economy, it is an absolute essential if a transportation system is to continue to function for any length of time; and a really serious stoppage at any one point in the system has a cumulative 'piling-up' effect throughout that system, including its terminals. Supplies have to be cleared from the source to enable production to continue, and so ships arrive and discharge more rapidly than the ports can be cleared; trains keep arriving at railheads and the empties cannot be sent back to the base, and so on. It is easy to create congestion and often extremely hard to clear it; and the result of all this is to form accumulations of rolling stock or transport which not only dislocate the working of the system, but in themselves form very conspicuous and vulnerable targets. The distance to which the repercussions of a single stoppage can extend is often astonishing. For instance, during the German concentration for the winter battle in Masuria in February 1915 a collision occurred near Könitz in East Prussia: the wreckage was rapidly cleared, and the resultant delay at the detraining stations not far distant was not very great; but one of the formations assembling for the battle was the XXIst Corps coming from the La Fère area in France via Sedan and Trier; and the result of this minor stoppage on the Russian front was a delay of thirteen hours in the Sedan district, hundreds of miles away, the other side of Germany. Moreover, it is not only an actual stoppage that dislocates traffic; it was the experience on both sides during the War that the mere threat of enemy bombers. being active, especially at night, caused serious delays and interruption of rail programmes by holding up trains and causing them to damp down fires when bombing was going on. This, then, is the point that it is important to understand—that it is upon dislocation, not destruction, that air action against communications relies for its effect.

RAILWAYS

'European warfare as it was practised on the French and Polish frontiers was a struggle for railways, conducted by men at the end of railways who would be reduced to fisticuffs in a week and starvation in a fortnight if their railways could be paralysed. . . . It was the railway and the railway alone that made possible the vast and paralysed armies that lay helplessly opposite each other across Europe.'[1]

Those days are already nearly twenty years behind us, and we cannot suppose that we shall ever again see such 'vast and paralysed armies' as existed, sometimes precariously, at the

[1] Philip Guedalla, *Men of War.*

end of railways in the years between 1914 and 1918. Perhaps one of the greatest revolutions of the many in the post-war era has been the rise of road transportation in civil life, a revolution which has been reflected in military practice. But the road has its very definite limits, and for the rapid carriage of commodities in bulk the railway still remains, and probably for many years will remain, supreme. And although road transport has already increased the flexibility of transportation, the day cannot yet be foreseen when the road could replace the railway as the most vital form of communication for great armies.

The history of railway bombing in the last War unfortunately does not contain many lessons of modern application. Indeed, it is mainly a history of failure to achieve decisive results, though at least one of the reasons for that failure is instructive. A fairly full description of the railway bombing undertaken before and during the major actions in France can be found in the Official History, and space will not permit of more than a brief reference in this book.

The battle of Neuve Chapelle in March 1915 was the first occasion on which a definite programme of railway bombing was undertaken; and there was a serious attempt to prevent the transfer of enemy reinforcements to the battle area, by bombing the stations at Courtrai, Menin, Lille, Douai, and Don; in the course of which a very gallant attack on Courtrai did result in traffic through that junction being delayed for about three days. Again, during the second battle of Ypres a number of points on the railways were bombed in an endeavour to prevent the movement of enemy reserves from Ghent to the Ypres Salient. The results on the whole were disappointing. An investigation held in July 1915 of the results of railway bombing up to date disclosed the fact that out of 141 attempts to hinder movement by bombing railway stations, only three had been successful.[1] A new policy was consequently initiated under which attacks on railway communications were only to be undertaken under orders from G.H.Q.; and the conclusion was reached that the best method of interfering with rail movement was not by bombing stations but by wrecking trains on the move, when possible in a cutting. Another important policy agreed on at this time was of co-ordination of the British

[1] See *The War in the Air*, vol. ii, p. 117.

and French air effort against enemy railways, in strict conjunction with important operations on the ground—a decision which, though it was given effect at the battle of Loos, was unfortunately not sufficiently fully implemented in the years that followed. For the Loos battle, plans for

'a comprehensive attack on the enemy rail communications to the battle area had been drawn up at conferences between the French and British Air Services. The French bombing squadrons on the British right were to attack the junctions at Cambrai, Roisel, Tergnier, and Mézières, and French airships were to make night raids on the junction at Busigny. To the Royal Flying Corps was allotted the important railway triangle Lille–Douai–Valenciennes. The latter junction was to receive special attention to hamper the movement of ammunition and reinforcements from the great depots at such places as Mons and Namur.'[1]

In spite of unfavourable weather a good deal of damage was done, particularly at Valenciennes where hits were obtained on ammunition trains, and the Germans admitted that all traffic was stopped for some time. In passing, a point that deserves comment is the nonchalance with which the Germans appear to have left ammunition trains lying in important junctions on main lines, such as Valenciennes, St. Quentin, and Thionville, in each of which places great damage was done at different times by ammunition trains being hit by our bombs.[2] It is almost a platitude to say to-day that ammunition railheads should be kept off main lines, and ammunition trains should not be allowed to stand in important junctions. The bombing during Loos did not, however, achieve any really important results, and the Germans say that although some inconvenience was caused, 'all units and formations called up to reinforce the front arrived at their destinations up to time'.

The end of 1915 saw the beginning of mass bombing by considerable formations. An instruction issued by Head-quarters, Royal Flying Corps, in December[3] stated that 'it is now an accepted principle that attacks on all important objectives should be carried out by as many aeroplanes as possible' and

[1] See ibid., vol. ii, p. 127.
[2] The Germans were not the only offenders. See ibid., vol. iv, p. 364, for the destruction of Lillers station from this same cause.
[3] See ibid., vol. ii, Appendix VI.

that action against trains in motion was to take the form of 'continuous bombardment by small detachments of aeroplanes'. This was followed in February 1916 by another instruction[1] which laid down that 'Damage to railways can be so quickly repaired that no appreciable results are gained by attacking them unless such enterprises are undertaken at the right moment, i.e. at a time when even a temporary interruption to traffic on these railways would interfere with important operations then in progress'. This policy was sound enough, though it could be carried too far.[2] And the instruction did not touch on the really important point that no results of any value can be expected unless the attacks are carried out in strength by every available aeroplane—other less vital targets being temporarily neglected—and persisted in continuously, long enough to cause real dislocation.

There can be no advantage in going farther into detail about the railway bombing in all the battles of the Great War. The plans for the battles of the Somme in 1916, of Arras, Ypres, and Cambrai in 1917, and for the defensive battles in the spring of 1918, all included more or less ambitious programmes on lines similar to those outlined above, of which the particulars can be found in the Official History. Enough has been said here to show that, although no really serious interference with enemy reinforcement or supply resulted, the subject was by no means neglected, even as early as the spring of 1915. But although the direct result of the railway bombing upon the land operations was not of much moment, it had an indirect influence of the highest importance—which curiously enough probably goes far to explain why the direct results were not more significant. The fact was that, odd though it may sound, interruption of rail traffic was not the primary object of the railway bombing. It has been explained earlier in this book that the last War was, from the air point of view, above all an 'Army Co-operation War'—and the primary end to which were directed the efforts of the bombers as well as of the fighters was the establishment of local air superiority over the zone in which the battle was in progress on the ground, to enable our artillery and reconnaissance machines to carry on their work of close co-operation with the army. So the first object of the railway bombing was

[1] See *The War in the Air*, vol. ii, p. 182.   [2] See p. 171 below.

to draw off enemy fighters from the battle zone; and this object it did on the whole achieve with marked success, notably during the battle of the Somme, with results which have been described in an earlier chapter. It was the importance attached (and no doubt at that time rightly attached) to this factor of air superiority over the battle line, that always militated against the concentration of bombers on railway objectives to a degree that could really effectively dislocate a railway system of such immense capacity as the complicated network of lines behind the German front in northern France. One finds, for instance, in the orders by Head-quarters, Royal Flying Corps, by the 5th Brigade and by the 9th Reserve Wing, for the third battle of Ypres,[1] that practically every single objective selected for bombardment was an enemy aerodrome—the only railway bombing ordered was to be done at night by F.E.s of No. 100 Squadron and by the Special Duty Flight. As a matter of fact it was on this Army front later in the year, in connexion with the fighting at Passchendaele, that a study was made of the problem of interference with the enemy railway system in Flanders. The Intelligence Staff examined in some detail the various vital stretches and vulnerable points on the system; gave the location of the most important depots and dumps, and pointed out that the important stretch of line between Roulers and Thourout was within reach of long-range artillery, and therefore need not engage the attentions of the air force. This was a very valuable appreciation; and though very likely other similar ones may have been produced, this was evidently not the usual custom. Written orders can be overdone; but there seems little doubt that had the system of personal discussions between army and air force commanders been more often supplemented by committing their conclusions to paper, the employment of the air force might often have been more effective than it was.

General Foch's directive of April 1st, 1918, has already been referred to in a previous chapter; actually one of his first actions on being appointed Generalissimo must have been to issue these instructions 'with a view to ensuring the co-operation of the British and French Air Services during the battle',[2] i.e. that which had opened about ten days previously with the great

---

[1] See ibid., vol. iv, Appendixes VI to VIII.
[2] See ibid., Appendix XVIII.

German offensive of March 21st towards Amiens. In these instructions he allotted reconnaissance areas and special reconnaissance tasks to the British and French air forces; laid down a policy for the employment of fighter aircraft; and made arrangements for the dissemination and interchange of intelligence between the two Allied air head-quarters. In the section dealing with Bombing he selected eight objectives, all of which were important junctions on the railways leading towards the Somme area, and allotted four each to the two Allied air forces. The British Air Staff affirmed at the time that they were already operating on the lines laid down in this directive. Actually this claim was hardly in accordance with the facts, especially in relation to that section of the directive which laid down that 'effort should not be dispersed against a large number of targets'. The British air programme of counter-preparation against the German offensive only about a fortnight previously had included something in the neighbourhood of fifty targets[1]—a rather different thing from the eight which Foch had allotted to all the Allied bombing forces in his instructions. Apart, however, from the interest of this directive as another example of the thorough understanding by Foch and his great Staff Officer Weygand, even in those early days, of the first principle of air warfare—that of concentration—the point to note here is that it was by attack on railways behind the front that Foch considered the bombers could most effectively assist in arresting the German advance on Amiens.

The next development of interest in connexion with railway bombing was the study of the subject already referred to in a previous chapter,[2] which was undertaken by General Nash and the transportation experts. Their suggestions were put forward early in April 1918, when there was apparently the slenderest margin between the Allied cause and disaster, and when the supreme preoccupation of every responsible authority on the Allied side was to hold up the terrific enemy offensive towards Paris and the Channel ports. Their advice, which was most valuable and of a type which an Air Staff should welcome from its technical advisers, included the selection of three sections of railway leading towards the Somme front on which complete

---

[1] See *Military Operations, France and Belgium, 1918*, p. 101: 'The operations were distributed over a large number of widely separated objectives.'   [2] See p. 77.

interruption could be effected by the bombardment of a minimum number of objectives; those points on the enemy system which appeared most suitable for attack, from a technical railway standpoint, being specified. It is true that these recommendations included some details with which the experts, with more experience and time for consideration, might not agree to-day; but on the whole they were excellent, and in any event represented a serious attempt, by a body of specialists, at a reasoned appreciation of the methods of air attack against railways. Curiously enough their reception by Air Headquarters was anything but enthusiastic. It has already been explained that the Air Staff did not consider that there were enough aircraft at their disposal to give effect to the experts' advice; but it was further claimed that Nash had told them nothing with which they had not long been familiar, and that he had no idea of the limitations of air bombardment. This last was probably quite true; but if, indeed, the Air Staff were already familiar with the factors explained by the experts, the fact was not very evident from their actions at the time. For instance, Nash had told them that, for various reasons, junctions were not the most profitable points for attack; yet the Air Staff were advocating the bombing of junctions—a curious reversal of the policy arrived at in the summer of 1915, due presumably to the results of experience in the intervening years. Again, Nash had advised against bombing junctions, for technical reasons connected with the practical operation of a railway. Yet the Air Staff were urging the bombing of junctions on the grounds that the military damage likely to result was greater than that which could be expected from attack on other targets. Of course the decision, or rather the responsibility for advising the Commander-in-Chief, on the objectives to be attacked must always rest with the Staff, after they have weighed up all the various factors affecting the particular problem concerned. But the Staff will usually be unwise to decide for or against any particular objective *in direct opposition* to the advice of their technical experts—unless for some technical reason again, connected, for instance, with the range of aircraft or the bombing accuracy to be expected. Conversely, the technical advisers should not encroach upon the province of the Staff. For instance, on the subject of air attack on the watering arrangements of a railway, the experts said that water-towers, as targets, were too small to give a reasonable prospect of successful attack. This was a matter for the Air Staff, not for the transportation experts, whose advice on this point should be limited to the effect, from a railway point of view, of the destruction of watering facilities and a description of the most vulnerable points in those facilities.

Ultimately, as the result of General Nash's persistence and of further consultation between the Staffs, three alternative schemes were drawn up, designed to hinder hostile concentration against three different sections of the British front. One was to deal with a threat on the front La Bassée–Ypres, another with the front Arras–Albert, and a third with the section from Albert to the junction with the French about Montdidier. Each scheme provided for attack on four specified railway objectives; and eight squadrons were allotted—four of day-bombers and four of night-bombers—to operate under the orders of the Air Officer Commanding at General Head-quarters exclusively against those targets. Eight bombing squadrons out of thirty (or twenty-five if the Independent Force be excluded) was not a very large proportion of our available resources to be concentrated against a menace on the scale represented by the German offensives of 1918; but it was a beginning. Actually, as it turned out, only one of these concentrated bombing schemes was ever put into effect, and that not fully; because very soon after they were issued the tide turned, and it was the Germans who were on the defensive for the remaining months of the War.

So on the whole the direct *results of railway bombing in France*, measured in terms of dislocation of enemy rail movement, were disappointingly meagre and disproportionate to the amount of effort involved. *The reasons* for this are not far to seek. To begin with, of course, our material and technique, even at the end of the War, were relatively primitive; in the early days of 1915 there was not even such a thing as a bomb sight in the Service, and bombing was done by the 'chuck and chance it' method, aided by various private arrangements of wires, and marks on the leading edge. The C.F.S. bomb-sight, considering the date of its origin, was a remarkably efficient instrument; but even that was not comparable to the instruments of to-day —any more than the bombing aircraft of the war years were comparable either in performance, as bombing platforms, or in the arrangements for aiming and releasing the bombs, to the highly developed bombers of modern times. The bombs themselves were relatively inefficient from a ballistic point of view, and not infrequently failed to explode. Then the training of the personnel was necessarily inadequate; we had to develop our methods as we went along, from trial and error in action; pilots had to go on active service with very little flying experience, and often with no training at all as bombers; there were no practice camps or bombing ranges, and the only bombing training that was possible had to be done over the camera obscura which, though very useful, was not a really effective substitute for live bombs. The constant interference with bombing operations by bad weather was also due largely to the lack of training and inexperience of the pilots, to the unreliability of engines, and to the absence of adequate navigational instruments such as turn-indicators; and weather conditions which to-day would not interfere with flying, and in some respects would even be a definite advantage to the bomber, in those days were enough to keep aircraft on the ground. But the most important reasons for the disappointing nature of the results obtained by railway bombing in the War were a failure to concentrate sufficient aircraft at the decisive point, and a tendency to disperse the efforts of the aircraft that were concentrated over too many targets, and for insufficient length of time. During the earlier years, of course, we had not enough bombers at our disposal in France to entitle us to expect very striking results. Later on, though we had the aircraft, the organization of the air force in the field on the basis of decentralization to armies made it difficult to effect a really formidable concentration, even had we sufficiently appreciated its importance. The influence in this respect of the requirements of air superiority over the battle zone has already been touched upon in this chapter; and the tendency to attach too much importance to direct attack on the enemy air forces is one which must always be guarded against. The one essential lesson which all air force commanders must constantly hold in the forefront of their minds is this principle of concentration. If we concentrate every available aeroplane at the decisive point, against the smallest possible number of targets, and for really adequate periods, then modern air bombardment, against railways or any other class of target, may well have decisive results.   But if we neglect this principle our bombers will be a nuisance to the enemy—probably a serious nuisance, as they often were to the Germans in France—but they will be nothing more.

The necessity for expert technical advice on the selection of objectives for bombing has already been emphasized several times in this book.   For no class of objective is this so necessary as for railways; and in the brief examination which follows of some of the methods of attack on railways the necessity will no doubt become apparent for the opinion, from a practical operating point of view, of the railway authorities who will always be available in the transportation services in war.

The first and most obvious *Objectives in a Railway System* which suggest themselves for attack are *the junctions and main-line stations.*   It has already been seen that the transportation experts in 1918 did not consider these to be the most suitable objectives, on the grounds that the large number of alternative lines through big stations made it practically impossible to cut them all (we cannot always count on the enemy being so obliging as to leave a loaded ammunition train standing in the station as he did at Thionville); that repair facilities and personnel are on the spot;[1] and that it was very little use causing a stoppage which could be cleared up in a couple of hours. These arguments are undoubtedly true, and the big main-line junctions are by no means always the most suitable points for attack; but it is too much to say that they never are.[2]   Each situation must be considered on its merits, and the decision will

[1] For instance, the damage to the railway triangle at Longeau, just outside Amiens, on the night 24th–25th of March, 1918, was repaired in under twelve hours. Even so General Plumer in a letter to G.H.Q. dated June 2nd, 1918, quoted in the Official Air History, cites the bombing of Longeau as an example of the dislocation that can be caused by bombing important railway centres. See *The War in the Air*, vol. iv, p. 318.

[2] As so often happens, the experts disagree about this—which is the best reason for the Staff Officer to have sufficient knowledge to be able to give a reasonable decision upon conflicting advice. An authoritative article in an Austrian military journal takes a view opposed to that of the Allied experts in 1918, and says that: 'The chief points of attack will on the contrary be vulnerable junctions and engineering works, all the more so because at these points interference and damage will cause the most serious interruption to traffic.'

depend, for one thing, largely upon the density of the railway
system under attack—in relation to the strength of the bomber
force available.   Thus in a country with a very intricate and
highly developed railway system, like that on either side of the
Franco-German frontier, it may be necessary to select as
objectives the focal points in that system, i.e. the junctions—
because an attempt to dislocate the system, for instance by
derailing trains on selected sections of line, might mean an
undue dispersion of effort over far too many targets.[1]   Again,
the effect of air bombardment on a junction station will
depend on the situation there at the time, the density of the
traffic passing through it and the consequent degree of dislocation caused even by a short stoppage in the flow of traffic.
Perhaps the supreme example of this was the junction at
Amiens in the early stages of the Somme battle, when nearly
three hundred trains a day passed through the station; bombing
on anything approaching a serious scale at that time could not
have failed greatly to have reduced that number, and created
tremendous disorganization in that part of the system at a very
critical moment.   Traffic of this intensity will admittedly be
rare.   But there were plenty of occasions—for instance during
the great strategic rail moves in East Prussia—on which delay
even of less than the two hours postulated by Nash as the minimum must have caused some dislocation.   The routes by which
the German 35th and 36th Divisions were concentrating before
Gumbinnen in August 1914 crossed each other at a junction
called Deutsche Eylau; and six months later during the concentration for the winter battle in Masuria those of the XLth
and XXXVIIIth Reserve Corps crossed over at Bromberg—
each operation involving a density of about sixty trains per day
through the junctions named.   The selection of this sort of
objective obviously would require early warning from Intelligence sources, amplified by air reconnaissance reports, of the
railheads and routes in use, if the necessary deductions were to
be made in time; and the Intelligence Service should also be in a
position to say whether the lines crossed each other by bridges or
by converging for short stretches or merely by cross-over points
within the main junction—which last would probably increase

[1] See, for instance, p. 178 below and Sketch-map 2 for the alternatives to attacking Laon and Cambrai junctions during the battle of Amiens.

the chances of dislocation under bombing. So a detailed knowledge of the lay-out of a junction is necessary to judge the degree
of mutual interference between various streams of traffic, and
to decide where in that lay-out are the most vulnerable points.
For instance, most big junctions have avoiding lines, and if there
are many lines through the station it will probably be as well
to try to make the blocks at the points where the avoiding lines
join the main lines on either side of the station. An actual example of successful air action against a big junction at a critical
moment was the raid (of which, apparently, Sir Frederick Sykes
does not approve[1]) by a squadron of the Independent Force on
Thionville in conjunction with the French attack on the Marne in
July 1918, when the squadron was lucky enough to find an ammunition train halted in a siding and blew up the whole station, completely blocking all movement through that most important rail
centre for about forty-eight hours. And soon afterwards, during
the American offensive at St. Mihiel, another attack was successful in stopping movement for twenty-four hours through the great
railway triangle at Metz Sablon. Examples of successful attacks
on junctions are unfortunately few, but they do serve to indicate
what might be expected from action of this sort on a really intensive scale, especially in conjunction with operations on the ground.

Nevertheless the conclusion arrived at in the summer of
1915, and again by the transportation experts three years later,
that complete interruption of traffic was more likely to be
achieved by making the breaches on open stretches of line away
from stations, is sound in the main. Incidentally it takes added
weight from the fact that really important junctions may often
be strongly defended, whereas it is quite impossible adequately
to defend hundreds of miles of line between stations. And the
best way of stopping traffic is not merely by blowing a hole in
the track—unless, of course, a bridge can be destroyed—but by
derailing a train, if possible in a cutting where it is much more
difficult to clear the wreckage and repair the line than on the
level or on an embankment, especially the latter, where the derailed coaches can simply be rolled down the bank and the metals
quickly patched. It must largely be a matter of luck to find
a train actually in a cutting, but trains as such are not a very
difficult target; and they are very conspicuous, by reason of the

[1] See p. 79.

steam by day which can always be seen for miles, and of the glow
from the firebox by night. Indeed, the problem of the anti-
aircraft defence of trains is a very difficult one, and one which
has not been given sufficient attention up to the present. It will
certainly be necessary to mount anti-aircraft automatics on all
troop trains, either in open trucks or on the roofs of coaches
near the engine; and special arrangements will have to be made
for supply trains. But this will not be effective at night, and it
is the hours of darkness that offer the most suitable opportunities
for the bomber. Indeed, the Air Staff in July 1918 made the
bold claim—which had considerable support from the facts—
that a train at night was completely at the mercy of the low-
flying bomber. It is worth considering whether in the years
since the War we have not unduly neglected that valuable class
of aeroplane the light night-bomber, exemplified by the old
black F.E. 2 B of the war years.

The electrification of railways will make traffic less conspicuous and possibly on the whole less vulnerable. But the
power-houses and transformer stations constitute most vulnerable and vital objectives, of which the destruction would
paralyse all movement over very large areas; and it is significant
that the French are doing no electrification of their railways
near the German frontier.

In selecting sections of line where blocks are to be made it is
necessary not only to look for features like cuttings or viaducts,
but also, if possible, to pick out stretches where traffic is likely
to be especially dense—a point on which some guidance can
be obtained from a survey by the technical experts of such
details as gradients and other factors affecting the capacity of
the line. It is hardly to be hoped that we shall ever in future
be presented with such an opportunity as the single by-pass line
round Liége between the 24th of August and 2nd of September,
1914: this line, on which gradients of one in thirty necessitated
four locomotives per train, two in front and two behind, was—
on account of the demolitions in the Liége area—the only line
of rail communication for the three northern armies of the
German right wing in Belgium during that vital period of ten
days. The German strategic moves in East Prussia again provide useful illustrations of this point;[1] there were several sections

[1] See Sketch-map 3.

between junctions on which strategic lines converged for a number of miles, such as the section between Thorn and Schönsee where the routes of the XXth and XLth Reserve Corps converged during the concentration for the Winter Battle; or a similar section between Dirschau and Marienburg—both sections carrying from sixty to seventy trains per day. The Koblenz–Trier line up the Mosel valley is another example of a section where important strategic routes converge, and was used by five Corps during the German concentration in the west in 1914.[1] It amounts to this: that, other things being equal, sections of line should be selected for attack where the capacity is greatest measured in terms of trains per day, and where in consequence the dislocating effect of any one stoppage will be most serious.

Next, having succeeded in making a block, whether by wrecking a station, blowing in a cutting or tunnel mouth, or derailing a train, we must endeavour to *keep that block closed*. Wreckage on a railway, even of a badly crashed train, can be cleared away surprisingly quickly, the permanent way repaired and the line reopened to traffic, provided the *repair train* can reach the scene and the *break-down gang* get to work quickly and unhindered. Thionville Station was completely flattened out, and it should have been possible by constant harassing, bombing by day and night, to prevent the junction being rendered fit for through traffic in as short a period as forty-eight hours. Immediately on receipt of the report of a successful attack on a railway, arrangements should be made for the constant harassing of the crash by relays of aircraft at the shortest possible intervals, including single aircraft at night, when break-down gangs will at least be severely handicapped by having to work without lights. A direct hit on a repair train, with its equipment of cranes and lifting gear, will of course be of the greatest value, because the number available is never unlimited. Incidentally, an uncomfortable thought is that of the possibility of the use by an unscrupulous enemy of a judicious mixture of some persistent chemical such as mustard gas with the H.E. bombs in attacks on railways. The extent to which this would increase the difficulties of the break-down gang is easy to imagine, and

[1] For further examples see the very interesting article by Major C. S. Napier, R.E., 'Strategic Movement by rail in 1914', *R.U.S.I. Journal*, Feb. 1935.

it is worth considering whether we should not be ready with arrangements for decontamination of wreckage, and gas-proof clothing for repair personnel within the bombing area in war.

A railway is a very intricate and complex organization, and an enormous volume of routine administrative work is required for the regulation of traffic, the compilation and adjustment of time-tables and programmes of engine working, the allotment, distribution, and loading of rolling stock and a hundred other details. Centralized control of train running means greatly increased operating efficiency; and so it is obvious that if the railway administration can be disorganized by bombing the *control offices* at important centres like the head-quarters of Railway Operating Departments, considerable dislocation is bound to result—if only by forcibly imposing the less efficient method of decentralized control. A District or Divisional Superintendent may control as much as a thousand miles of line—Köln District Office during the British Occupation, for instance, controlled about 1,700 kilometres of track and 800 stations of various sizes. And the extraordinary efficiency with which the vast strategic train-moves on the German eastern front in the autumn and winter of 1914–15 were organized and controlled, was due to the foresight and excellent organization of the railway service, with its director, known as D. R. Ost, at the Head-quarters of the Eastern Army, working through District Superintendents at Königsberg and Danzig. Constant adjustments and alterations of schedules, diversions of traffic and instant arrangements for repair, are essential to meet the ordinary accidents which arise from natural and technical causes when a railway is working at really high pressure in war. This sort of work obviously will be vastly increased if the railway is also subject to systematic interference from the air at a critical period; and if at the same time the brain of the system in the control offices can be attacked, it may surely become impossible to retain control, and irretrievable confusion may result.

The uninterrupted working of an efficient *system of intercommunication,* telephone or telegraph, is, of course, an essential feature of the system of centralized control; and this, with the electrically operated *signal apparatus,* is another objective on which successful attack would inevitably result in widespread dislocation. Already at some big termini the signalmen in subsidiary signal-boxes are being replaced by automatic remote control from a central box. The system of controlling train movement known as 'absolute block', which is normal in most continental railway systems except for certain sections of the État line in France, again depends on uninterrupted communication. And the cumulative effect of a stoppage at any one point is well exemplified by this system, which guarantees a safe interval between trains by ensuring that no train can be allowed into a section of line until the next ahead has been signalled out of that section; so that a failure to signal a train out of a section—whether because the train has been crashed or because the inter-communication system has been cut— automatically results in a hold-up all along the line. The terrible collision on the French railway at Lagny in 1933 is an example of the possible result of a combination of a derailment of a train and the cutting of the electrical signalling gear, on a system of absolute block. Overloading of telephone lines was a fruitful cause of difficulty during the French concentration in 1914, and was a contributory cause of at least one of the few cases of serious congestion in East Prussia—that which occurred during the German advance to the Vistula in October 1914. The results of the bombing of the Turkish telephone exchanges during the final offensive in Palestine have already been noted, and there is no reason to suppose that air action cannot at least multiply the difficulties and delays due to overloading of the railway telephone and telegraph system.

If the brain of a railway system is the Control Office; if the nerves are the telephones and signal-apparatus and the arteries are the main lines; then the heart is represented by the *marshalling yards* where the bulk trains of different commodities are shunted and sorted, and supply trains made up for dispatch to railheads serving the various field formations. It may be that these are almost the most vulnerable points on the railway, and a stoppage or disorganization of work in a marshalling yard would have rapid and widespread repercussions. Signalmen, shunters, and engine drivers would have to be very well disciplined and unemotional to carry on under frequent harassing bombardment, and a hit on the signal-box or gear operating the points would bring work to a standstill. Moreover, a great deal of this sort of work has to be done at night, and the mere threat of bombers reported in the vicinity would at least cause serious delays by plunging the yards into darkness.

Bound up with the question of attack on junction stations and large railway depots is that of the extent to which bombing of *engine sheds and workshops* is likely to be worth while. The view of the experts during the War was that it was not likely to be profitable, in view of the enormous resources in workshops at the disposal of the enemy at the time, and the relatively slight material damage that could be inflicted. The circumstances in which this view was expressed were those in which it was urgently necessary for any action taken to produce *instantaneous* reactions on the situation in the battle line—and also when the volume, accuracy, and general efficiency of the air action available was not very formidable. In such conditions there is no doubt the conclusion was sound. But the experts would probably revise their views if they were asked to consider the possible results of air action on a modern scale, not as an occasional episode, but as a constant feature in a co-ordinated programme of attrition against a railway system during a period of preparation for intense activity on the ground. Furthermore, they were attaching, perhaps, disproportionate importance to the aspect of material damage, and not enough to the moral effect— measured in terms of disorganization of labour and restriction of output in these very large and conspicuous workshops. No doubt many belligerents will enlist their railway personnel on mobilization and subject them to military discipline and law; but the hastily conscripted workman, pursuing his normal avocation under arduous and dangerous conditions, cannot be expected to stand up to punishment to the same extent as the trained soldier in a fighting unit. Dislocation and restriction of output are bound to ensue, even from a mere warning that enemy bombers are in the vicinity—especially in shops which have once been heavily bombed. And properly directed air action against the depot workshops of a railway can hardly fail to affect adversely the work of maintenance, repair, conversion, and manufacture that is so necessary to compete with the shortage of *rolling stock* which is inevitable, particularly in the earlier stages of a war. As a matter of fact, as late as 1918 the

supply of rolling stock was a weak link in the German administrative system, and the serious coal crisis that arose was due not
so much to an actual shortage of coal as to a shortage of rolling
stock to lift it.

Again, it was no doubt true that on the western front in 1918
the number of *locomotives* that could be destroyed by the air
effort then available would not, as the transportation experts
pointed out, have any appreciable effect on the enemy's
transportation facilities, because any engines we could destroy
could only form an insignificant percentage of the total number
at his disposal.   But similar considerations to those outlined
above apply equally to this question of locomotives; and the
locomotive situation in other theatres of war, and at other times,
was not by any means so favourable as that in France in 1918.
The Russians, for instance, were by no means well off for locomotives, and the difficulties which would have resulted from the
loss even of a comparatively small number would have been
enhanced by the existence of two different gauges in the Polish
Salient.   In Austria there was a good example of how detailed
railway Intelligence, properly used, might assist in the concentration of air action on a particularly profitable objective
with the fullest economy of effort.   One of the limiting factors
in Austrian troop concentrations against the Russians and the
Italians was a shortage of special high-powered locomotives
capable of hauling fully loaded military trains up the steep
gradients in the Carpathians and the foot-hills of the Alps.   Now
it is not impossible that these large engines may have been
housed in special sheds or depots for maintenance or repair
when not actually in use; and if this were the case then those
depots would obviously have been a very important objective,
because the loss of a number of the special locomotives would
have thrown a heavy strain on the Austrian resources in others
of a less powerful type, of which four were required to do the
work of one of the heavy engines.   Of course these special sheds
or depots *may* not have been necessary at all, but the story is
quoted as a useful illustration of the value of good Intelligence
in conjunction with expert technical advice.   Moreover, very
serious dislocation of traffic over limited periods—but long
enough to be critical if inflicted at the right moment—can be
caused without the actual destruction of locomotives.   The

Second Army Staff were very anxious about the 'round-house' turntable at St. Omer on which, as they pointed out, a hit by a single bomb might quite likely lock up forty engines for a period of several days. These large circular sheds, in which a number of locomotives are housed standing on a power-operated turntable, are not uncommon on the continent; and it is true that a single bomb on the turntable machinery would result in locking up for a considerable period all the locomotives except those which happened to be opposite the exit at the time. This actually happened in the raid on Metz Sablon already referred to, when ninety-four locomotives were rendered inoperative for several days until the turntable machinery could be repaired. Incidentally, this was one of the measures of passive resistance adopted by the Germans on the occasion of the French occupation of the Ruhr; on the first day about 170 locomotives were locked up by sabotage in the round-house loco sheds.

One or more of the various features of a railway system mentioned in the foregoing will usually be the most profitable objectives for attack, and the objective, or combination of objectives, selected will depend upon the conditions at the time, such as the effect it is desired to produce, the density of the system under attack, and the efficacy of the hostile defences both in aircraft and artillery. The Intelligence Staff or the technical experts may be able to suggest others which at certain times or in certain countries may be very important. For instance, it may be possible by mixing incendiary with the high-explosive bombs appreciably to reduce the enemy's *fuel reserves*, and thus interfere with traffic, by bombing coal dumps—the Turkish railways serving the Palestine and Mesopotamian fronts were often hard up for coal, which had to be brought from a great distance and stored in dumps at certain centres. Or, especially in a very dry country, a very vulnerable spot on a railway might be the cisterns and *watering arrangements*. Again, there may exist within range on important main lines certain large stations, such as are found sometimes on the Continent, with roofs supported by large girders extending right across the lines; and a heavy bomb with great blasting effect bursting between the walls may blow them outwards, dropping the roof across the rails and blocking the line for days.

No doubt there may be other objectives, and there are certainly a variety of factors other than those mentioned in this chapter which complicate and increase the possibility of congestion on the railways in war. For instance, there will be an enormous volume of traffic other than troop and supply trains which cannot possibly be banished from the railways—such as that necessary for the feeding of the civil population and for the distribution of essential commodities; for the increased demands of industry due to the mechanization of war; for the evacuation of refugees from threatened districts, or of prisoners after a battle; for the movement of reservists joining their units on mobilization, and of course (a point not always remembered) for the normal return of empty rolling-stock. But enough has been said to indicate the sort of points which have to be looked for in framing a plan for attack on an enemy's railways. Once more the point to remember is this—that whatever the objective, or class of objective, selected we must concentrate upon it the highest possible proportion of our available strength, and continue to do so for sufficient length of time to give a reasonable chance of decisive effect.

ROADS

Although the railway has not yet been seriously challenged as the best means of transportation in bulk over long distances, and will no doubt remain for many years the most important element in the line of communications of an army, still the enormous increase in the volume and capacity of road transport in recent years has already profoundly affected the problem of military transportation in war. In fact, movement to-day must be considered as a whole. Motor transport is sure to supplement and ease the burden on the railways to an increasing degree as time goes on; and by relieving congestion on the railways in periods of stress, will render them correspondingly less vulnerable and susceptible to dislocation from the air. It may even be that we have reached the peak of the air's capacity to interfere with rail movement, and that from now on it will remain static—advances in the efficiency of air action being offset by the increasing capacity of the road to relieve the railway—although there is a sort of saturation-point beyond which increase in road usage is liable to defeat itself; it is too early to be able to see clearly in that direction.

For the present, however, it is at any rate clear that the importance of the roads in the scheme of military communications varies in inverse ratio to the density of the railway system in the theatre of war. In undeveloped countries where railways are scarce a correspondingly large proportion of troop movement and supply traffic—and consequently of suitable bombing objectives—will be found on the roads. The extreme case is reached where there are no railways at all—such as across the Indian Frontier where, for example, an army operating beyond the Khyber, where the railway stops at Landi Khana, would be entirely dependent upon a line of communication consisting of a single road; and that lying for long stretches in deep defiles which would constitute almost ideal objectives for bombers. The line of communication of an army operating under these conditions may even constitute one of those special occasions referred to in a previous chapter[1] when the requirements of security become temporarily paramount as far as the air is concerned, and demand the first attention of our air striking force—even to the extent of providing close fighter defence; because we could not possibly afford to risk having that road cut for any length of time.

Civilized countries with a highly developed railway system will be well provided also with roads, and the problem of interference with road movement becomes a good deal more difficult. It is much harder to make a breach or block in a road than it is to do the same in a railway; and a breach when made can usually be cleared and made temporarily fit for the passage of traffic more quickly than on a railway. The destruction of a bridge will, of course, be the most effective and complete way of breaking a road; but that is not easy, and even then a temporary bridge fit to carry motor transport can be constructed far more quickly than a railway bridge. The most likely places to block roads will be found in towns or villages, where a heavy bomb in a narrow street will not only crater the road but by its concussion may blow in the houses on either side, and block the road with debris. And those who remember the condition of Holborn for some months in 1929 will be able to imagine the widespread and enduring effects of armour-piercing delay-action bombs getting through into the gas-mains in big towns.

[1] See p. 28 above.

As a rule, however, air action against roads must rely for its effect not upon actual damage to the roads, but upon stopping or dislocating the traffic using them; and for this purpose also the most suitable places will often be the bottle-necks and defiles, either natural or those formed by towns or bridges. The subject of attack on columns of fighting troops and transport on the roads has already been dealt with in the last chapter and need not be further elaborated here. Attack on supply on the roads forward of railheads will not normally be worth while; once supplies or ammunition reach the lorries of the maintenance companies or their equivalent, they become too scattered to constitute a profitable objective. There may be exceptions to this rule, when sustained attack on a road channel of supply during a period of active operations may be highly dangerous —especially in conjunction with a programme of railway bombing. One such exception was the famous Bar-le-Duc road into Verdun, which, with one light railway, constituted the only line of supply into the fortress during the siege. Accounts of the flow of traffic along this road vary from 2,000 to 6,000 vehicles a day, but it averaged something like 90,000 men and 50,000 tons of material per week for a period of seven months. And it is difficult to believe that Verdun could have held out in the face of continuous, concentrated air attack on this road, certainly on a modern scale. This again, however, is an instance of what really amounted to a large army dependent virtually upon a single line of road communication; and save in such exceptional conditions, bombers will usually be more suitably employed against objectives other than supply by road.

BRIDGES

Bridges as targets are superficially attractive, and are popular in hastily considered plans for air bombardment in staff exercises and schemes. In point of fact they are not nearly such suitable objectives as they appear at first sight. To begin with the really important bridges are almost bound to be defended— a published report by the head of the Chemin de Fer de l'Est even proposed that certain key bridges in the French mobilization scheme should be provided with balloon aprons against low bombing attack; and this will inevitably reduce the standard of accuracy against what are already very small targets. A really heavy bridge is almost impossible to destroy from the air, except possibly by a direct hit from a super-heavy bomb; and even that would probably only cut one section of the trackway and would not do much damage to the more important parts, the piers and abutments. It is believed that a bridge like the Hohenzollern bridge at Köln would require many tons of explosive, carefully placed, to effect anything like thorough demolition. The ordinary stone or brick arch bridge can fairly soon be made temporarily fit for use, for anything but very heavy traffic, either by bridging the gap or shoring up the arches with timber supports; while small ones, like that over the Luce at Hangard, quoted in a later chapter, can be made passable in a matter of hours. On the other hand, examples may arise of a bridge which strategically is of the highest order of importance, and yet is not of very solid construction. One such was the railway bridge which carried the only line from Berlin to Constantinople and the Peninsula across the River Maritza at Kuleli Burgas, and of which the destruction would have meant a serious breach in the only rail line of communication by which the Turks could get their sorely needed munitions from Germany. This bridge was within range even of the rather antediluvian British aircraft then available; and in fact was bombed several times, by landplanes from Imbros and float-planes from the carrier *Ben-My-Chree*—unfortunately without success, the nearest thing being a near miss by two 112-lb. bombs which delayed traffic for forty-eight hours by straining a pier. The strategical results of the destruction of a bridge of this sort would obviously be so vital as to justify the continuous expenditure of the effort of an adequate force of bombers, to make the breach and prevent its being repaired when made.

As a rule, however, it is better not to attempt to demolish bridges, but to regard them as defiles and attack the troops or transport that have to cross them—for which the technique and the type of bombs used is rather different. Bridges are very often the best example of that well-worn military expression the bottle-neck, and offer supreme opportunities for the dislocation of movement. For instance, during the first sixteen days of the War, from the 2nd to the 18th of August 1914, the two railway bridges at Köln carried a total of about 180 loaded troop trains moving west per day—representing a total of well

L

over a quarter of the total German concentration against France; a west-bound train crossed the North, or Hohenzollern bridge, every ten minutes for the first fortnight. These bridges were well within range of modern bombers from French bases, and the scale of attack that could be put down to-day—had it existed in 1914—could hardly have failed, even in the face of the inevitably strong defences, to have wrought havoc with that marvellous programme of complicated train movement, worked out in the minutest detail in peace, on which the great concentration depended. There will probably never again be an opportunity fraught with such vast strategical potentialities, but the tactical results of air action against road bridges in conjunction with operations on the ground may be of no mean value. A German report of the fighting on the Marne in July 1918 says that air action and artillery fire on the bridges 'dried up supply to the Divisions south of the Aisne'; and a really heavy concentration of assault aircraft on the 8th of August against the more important bridges over the Somme might have been the best way of stopping the flow of enemy reserves by road to the battle front on the first day of the Fourth Army's attack.

> 'The air war of the future, with targets far behind the front, or frontiers deep in the interior, creates a new standard of judgement. Hitherto we have been accustomed to regard the smooth handling of railway traffic as dependent only upon technical efficiency and a proper regulation of demands: the capacity of road communications has been reckoned simply from their density, standard of construction, gradients and so on, and waterways judged only by their navigability. But the proved results of air attack call in question all such assessments of value.'[1]

It is always difficult to adjust the mind to new standards of judgement, and the more so when some of the factors on which that standard must be based are still to some extent a matter of conjecture. In the foregoing chapters the writer can claim to have done little more than scratch the surface of a subject which obviously offers an immense field for thoughtful and unprejudiced investigation—a subject, furthermore, which im-

---

[1] 'Communications and their Susceptibility to Air Attack', article in the Austrian journal *Militärwissenschaftliche Mitteilungen*, November 1933.

periously demands the attention not only of the airman, from the point of view of attack; or of the administrator and transportation officer, to enable them to make the necessary adjustments in the system of maintenance and supply; but perhaps above all of the General Staff officer, because it raises issues of fundamental importance touching the whole field of military operations and organization.

PART III

THE BATTLE OF AMIENS, AUGUST 8TH–11TH, 1918

VIII

THE STORY OF THE BATTLE[1]

THE SITUATION WHEN THE PLAN WAS CONCEIVED

ON the 17th of July 1918 Sir Henry Rawlinson, then commanding the Fourth British Army, was in his head-quarters at
Flixecourt drafting a letter to Sir Douglas Haig, outlining and
asking approval for a plan to attack with his army astride the
Somme, with the object of disengaging the important strategical
centre of Amiens. As he wrote, the waves of the last great
German attack of the War were beating against the French
defences south and east of Rheims; and Mangin's divisions
were assembling in the woods between Compiègne and Villers
Cotterets for the counter-stroke that at dawn next day was to
mark the turn of the tide in the west. At the northern end of
the allied line there were strong indications that yet another
German offensive was imminent in the Lys sector, and in fact
we now know that Ludendorff was staging an attack in that
area for July 20th, though it was never launched, owing to the
pressure of Allied counter-offensives on other parts of the front.

Two other factors in the general situation are relevant to this
story. In the air the German aeroplane attacks on London had
been finally abandoned in the face of the very powerful and
highly organized system of passive defence; and although we
had still to provide against the possibility of a renewal of these
attacks, in fact no raid had crossed the British coast since May
19th. British counter-measures against German war industry
had been initiated during the previous winter, and by July
1918 five bombing squadrons, the nucleus of the great Inter-
Allied bombing force of the 1919 programme, were operating against the Rhineland from the Nancy area.[2]

[1] See Map North-West Europe 1/250,000 G.S. G.S. 2733 Sheet 4, and Sketch-
map 2 at the end of this volume.

[2] A sixth squadron, No. 97, joined the force on August 9th, from England.

At sea the enemy submarine offensive had been definitely checked during the latter half of 1917 by the adoption of convoy and the arrival of the American flotillas; and the menace had been still further diminished in the spring of 1918 by the mine-barrages and the operations against the submarine bases on the Belgian coast.

THE PLAN

In accordance with directions issued by Sir Douglas Haig at an Army Commanders' conference earlier in the month, the Commander of the Fourth Army on the 17th of July submitted his original plan for approval. This plan in its initial form was of limited scope, and aimed at important but definitely restricted strategical results. Its objects were to assure the safety of Amiens; to improve our position at its junction with the French, and from the point of view of ground for observation and defence; to shorten the line, and to strike a blow at the enemy morale, which was rightly assumed to be low after the failure of his Champagne offensive. The essentials on which the plan was based were firstly surprise and therefore secrecy; and secondly the method of attack by masses of tanks without artillery preparation, on the lines that had proved so effective and economical in life at Cambrai in the previous November, and more recently on a smaller scale in the Australian attack at Hamel early in July. (General Mangin was to give a further demonstration of the efficacy of these tactics at dawn the following morning on the Marne.) It was largely in order to make more sure of secrecy that General Rawlinson proposed that the attack should be made by the British alone, without the active co-operation of the French on his right.

The scope of this plan received its first enlargement when Sir Douglas Haig—who was being pressed by General Foch to undertake an offensive in the north—extended it to embrace the active co-operation of the First French Army (General Débeney) which was placed by the Generalissimo under his orders for this operation. On July 29th Generals Rawlinson and Débeney were told that the object of the offensive was to disengage Amiens and the railway Amiens–Paris; and that when the line Méricourt–Hangest was secured the British Fourth Army keeping their left flank on the Somme were to press the enemy in the direction of Chaulnes, while the French with their right on the Avre were to push on in the direction of Roye.

The aim of the operation was still further extended on August 5th as the result of a direction by Foch that initial success was to be exploited by pushing on towards Ham. On that day at a conference of Army Commanders at Fourth Army Head-quarters Haig indicated that in consequence of the wider de-velopment of the impending offensive three British divisions were to be concentrated in General Reserve behind the battle front, and others were to be held in readiness behind other sectors of the front, to be moved south if necessary. He gave directions—afterwards confirmed in orders—that the first objective would be the line Hangest–Harbonnières; the next move to be a push forward in the general direction Roye–Chaulnes '*with the least possible delay*',[1] thrusting the enemy back with determination in the general direction of Ham, and so facilitating the operations of the French from the front Noyon–Montdidier.

The development of the plan is described in some detail in order to emphasize that in the month between its conception and its execution it had increased in scope from an operation merely to disengage Amiens into one embracing a far more ambitious plan, holding out the most far-reaching results, and aiming at a penetration little inferior in depth and importance to the great German break-through on the Fifth Army front in the previous March. It is as well to bear this in mind when surveying the results of the attack, which in fact—after a most completely successful and apparently overwhelming initial break-in—was brought to a standstill on the fourth day on a line little in advance of that originally intended as the final objective in Rawlinson's original plan. For at this date the method which has been described as 'a series of rapid blows at different points, each broken off as the initial impetus waned, each so aimed as to pave the way for the next, and all close enough in time and space to react on each other',[2] had not yet crystallized into the definite strategical policy which so effec-tively wore down the power of resistance of the German armies in the last three months of the War.

[1] See *The Story of the Fourth Army*, p. 18.   [2] Liddell Hart, *The Real War*, p. 401.

PRELIMINARY ARRANGEMENTS AND ORDERS

General approval having been obtained from the Commander-in-Chief, the Fourth Army commander held a series of conferences to settle the detailed arrangements for the attack. Owing to the importance of keeping the project absolutely secret these conferences were held at a number of different head-quarters, and were attended only by the minimum number of senior officers by whom it was necessary that the development of the plan in its earlier stages should be worked out. Conferences were held on the 21st, 25th, 27th, and 29th of July, at the last of which Brigadier-General L. E. O. Charlton, commanding the 5th Brigade R.A.F. under the orders of the Fourth Army, was present for the first time. At none of these conferences was the share of the R.A.F. in the forthcoming battle discussed in any detail. Certain arrangements of minor importance were made, mainly concerning secrecy and inter-communication. And arrangements were discussed for aircraft to fly over the line at night to accustom the enemy to their noise, in preparation for the night of Z—1 day when that noise was to be used to cover the assembly of the tanks. At the conference on the 27th it was also mentioned that the 9th Brigade R.A.F. would join the Fourth Army, which as a matter of fact was not strictly correct: the 9th (G.H.Q.) Brigade did support the Fourth Army attack but never came under the orders of the Fourth Army commander. At a further conference held at the head-quarters of one of the tank brigades on July 30th— at which the A.O.C. was *not* present—the question was raised of air support for the tanks. The Tank Corps commanders asked that aircraft might be detailed to protect tanks advancing on distant objectives, by engaging enemy artillery with bombs and machine-gun fire; they were told to make these arrangements direct with the R.A.F.

It is, perhaps, hardly necessary to point out to-day that if the R.A.F. are really to understand what is required of them, and pull their full weight in battle, Air Force commanders must be taken into the General's confidence, and must attend the conferences at which the plans are discussed.

The battle orders issued at the end of July to the Corps Commanders and the Air Officer Commanding the 5th Brigade R.A.F. were amplified at a conference at the IIIrd Corps head-quarters on August 1st, at which the Army Commander described the scope of the operations as being to capture the old outer Amiens defences and to organize for defence the line reached as early as possible.

On the 6th—as the result of Foch's decision to extend the scope of the plan—these orders were amended, and at a conference at Army head-quarters that day the general objective given was from the point of junction with the French—Damery–Lihons–Méricourt–Étinehem–Dernancourt, which involved a penetration at its deepest point of about ten miles.

The attack, which was to be launched between the road Amiens–Roye on the south and the River Ancre on the north, a frontage of about 30,000 yards, was to be carried out by the Canadian, Australian, and IIIrd Corps in that order from the right. The final objective for the first day was to be the line of the old Amiens outer defences, a line—marked in blue on the battle maps and referred to as the blue line—some 3,000 to 5,000 yards short of the general objective allotted at the conference on the 6th. To achieve surprise and to economize in man-power the initial assault was to be done by masses of tanks, with comparatively few infantry in support; and for this purpose ten battalions of heavy tanks (Mark V and V*) were allotted, the 2nd, 8th, 13th, and 17th to the Australian Corps, the 1st, 4th, 5th, and 14th to the Canadians, the 10th to the IIIrd Corps, and the 9th to Army Reserve, making a total of 360 heavy tanks. There was to be no artillery preparation, and the task of the guns was to cover the assault of the tanks and infantry in the initial stages by barrage, to keep down the fire of hostile artillery by concentrations of counter-battery fire, and to search probable assembly areas of enemy reserves. The Cavalry Corps was placed under the Fourth Army for the battle, its task being to assist the infantry on to their objective, and subsequently to explore in the general direction of Roye and Chaulnes with a view to cutting the enemy's communications and assisting the French. The 3rd and 6th tank battalions (a total of 96 whippet tanks) were allotted to assist the cavalry.

The First French Army, which extended the attack to the right and was directed on Roye, had no tanks and was therefore to attack forty minutes after British zero, to allow time for a hurricane artillery bombardment by way of preparation.

The subsequent instructions to the various arms and services for their participation in the battle went into considerable detail. Those to the R.A.F.[1] allotted the six Army Co-operation Squadrons of the 15th Wing (5th Brigade), on the scale of one per Corps—No. 35 Squadron to the IIIrd Corps, No. 5 to the Canadian, No. 3 (Australian) to the Australian Corps, and No. 6 Squadron to the cavalry. No. 8 Squadron was detailed to work with the tanks; while to No. 9 was allotted the task of keeping the machine-guns of the IIIrd and Australian Corps supplied with ammunition on the second and third objectives— which was done by dropping the ordinary S.A.A. boxes, containing 1,000 rounds, by specially designed parachutes. Special markings to identify the aircraft engaged in these various tasks were laid down.

The eight single-seater fighter squadrons of the 22nd Wing were to be employed exclusively in low-flying attack against objectives on the ground. They were allotted in even distribution to each Corps front of attack, though this disposition was liable to be changed if it became necessary as the battle progressed. For the first four hours after zero they were to operate in the area beyond the first bound of the attack on the ground, about 3,500 yards from the starting-line; being subsequently 'lifted' to the area beyond the red line, about 5,000 yards farther east—with the proviso that very favourable targets were to be attacked wherever they were seen.

The ring was to be held by high flying patrols over the front of attack, from the six single-seater fighter squadrons of the 9th Brigade, which was the G.H.Q. Reserve Brigade of the R.A.F., and was used as a mobile reserve to reinforce any portion of the front where important active operations were in progress. These were to be reinforced if necessary by fighters of the 3rd Brigade—presumably by arrangement with the Third Army, on the immediate left, to whom this brigade belonged.

Finally it was mentioned that the two bomber squadrons and the fighter reconnaissance squadron (Bristol Fighters) of the 22nd Wing would be available; and that four day-bomber and three night-bomber squadrons of other Wings would co-operate. In the event this number was exceeded, and a total of eight day-bomber and five night-bomber squadrons took part in the

[1] See *The Story of the Fourth Army*, pp. 24–5.

operations on the Amiens front between the 8th and 11th of August. Oddly enough, nothing was laid down as to how the bombers were to be employed; but Sir A. Montgomery Massingberd, who was the Chief of Staff of the Fourth Army at the time, says in his book that 'the objectives of the day and night bombers were the railway centres at Chaulnes, Roye, Nesle, and Péronne, the crossings of the Somme, roads and billeting areas which the enemy were likely to use'. No doubt these details were arranged verbally between the General Staff and the Commander of the 5th Brigade.

THE BATTLE

It is proposed to examine the air plan, the orders issued by air force commanders, and the way in which these orders were carried out, in the form of a detailed commentary. And, in order to provide a background for this commentary against which the actual effects of the air action on the battle may be seen in proper perspective, it is convenient at this stage to describe very shortly the course of the battle, entering only into such detail as is relevant to the subsequent examination of the air operations. Attention will be confined to the four days August 8th to 11th inclusive, for reasons which will become apparent.

August 8th. The R.A.F. may justly claim some share of the credit for the remarkable success with which the preliminary concentration was concealed from the enemy. In the first week of August the Fourth Army was reinforced by five divisions, two cavalry divisions, nine tank battalions, and an additional thousand guns. This involved nearly 300 special trains for personnel, guns, and ammunition being run into the Army area, in addition to the ordinary supply trains. On the nights of the 6th and 7th the tanks were moved up to their preparatory and assembly positions under cover of the noise of low-flying aircraft; and when over 300 tanks rolled forward under the barrage in the misty dawn of the 8th of August, they took the enemy completely by surprise. The front attacked was held by eight weak divisions of Von der Marwitz's Second German Army, and so complete was the surprise that two divisional reliefs were actually in progress when the blow fell. The initial surprise was assisted by heavy ground mist in the early morning, which did not clear till nearly 10 o'clock. But this one advantage of the mist was heavily counterbalanced by disadvantages in the air, where the weather conditions seriously restricted the work of the close reconnaissance aircraft, interfered with observation, and hampered the operations of the low-flying fighters. Nevertheless the attack went according to plan, and—except for the subsidiary attack by the IIIrd Corps in the more difficult country north of the Somme, which failed to capture all the ground intended—was completely successful.

'On the evening of August 8th the situation on the Fourth Army front was most satisfactory. The main attack south of the Somme had been successful almost beyond the most sanguine expectations, and the Canadian and Australian Corps had reached their final objectives except for a small portion on their extreme Northern and Southern flanks. The losses of these two Corps had been exceptionally light, the largest capture of prisoners and guns taken on any one day during the war on the Western front had been made, and in addition the enemy's troops were thoroughly demoralized. Prisoners from eleven different divisions had been captured by the Fourth Army, there were few hostile reserves immediately available, and the prospects of further success on the following day were extremely bright.'[1]

On the right of the British the First French Army had made valuable progress, capturing Moreuil and Plessier, together with over 3,000 prisoners and a large number of guns.

The weather conditions already mentioned restricted air operations in the opening stages of the attack; but by 10 o'clock the mist had cleared and the squadrons were fully engaged. Very valuable work was done by the corps squadrons operating on the lines laid down in the instruction already quoted. The only feature of their work which calls for special comment was that the squadrons of the Canadian and Australian Corps laid smoke screens at selected localities by dropping phosphorus bombs with contact fuses; unfortunately no details can be discovered about such points as the object of the screens—presumably to cover the tanks—or the arrangements for ensuring that they were put down at the right time and place, which might have held valuable lessons for application to modern conditions of air co-operation with armoured forces.

[1] *The Story of the Fourth Army*, p. 51.

Apart from this, however, and the use of No. 9 Squadron as a sort of mobile S.A.A. reserve for the IIIrd and Anzac Corps, the work of the Corps squadrons throughout the battle was on normal lines, i.e. contact patrol and co-operation with the artillery. And if their activities do not receive further notice in this story, it is not because they were not useful but because they do not provide any lessons which are not already well known, or which it is the object of these chapters to examine.

The assault aircraft of the 5th Brigade operated in sectors corresponding to the Corps fronts in the area west of the Somme, two squadrons (less one flight) on the front of IIIrd Corps, three squadrons on the front of the Australian Corps, and three on that of the Canadians, while one, No. 65, was detailed to the Cavalry Corps. The conditions after the first break-through were well suited to this form of air action, approximating as they did to those of open warfare, against an enemy already shaken and to some extent demoralized by the suddenness and overwhelming nature of the British attack. Sir A. Montgomery Massingberd says: 'Flying very low . . . our aeroplanes completed the demoralization of the enemy by attacking his retiring troops and transport with bombs and machine-gun fire, and by shooting gun teams in the act of withdrawing the guns.'[1]

Meanwhile the ring was held by the high offensive patrols of the 9th and 3rd Brigades, who also co-operated with the bombers in their attacks on the enemy communications behind the actual battle area.

Eight day-bomber and three night-bomber squadrons, drawn from five different brigades, took part in the battle on the 8th and during the night 8th–9th. It had been appreciated that during the 8th, at any rate until some 10–12 hours after the attack began, bombing of railway communications was unlikely to strike reinforcements and reserves moving by rail to the threatened front. It was, however, considered important to restrict the enemy's air activity, and for this reason the enemy aerodromes at Moislains, Estrées, St. Christ, and Bouvincourt were bombed in the morning. But by far the greatest proportion of the effort of both day and night bombers was directed against the crossings of the Somme. During the first twenty-four hours after zero eleven bridges over that river from

[1] *The Story of the Fourth Army*, p. 50.

Pithon to Cappy were attacked from varying heights by a total of 162 individual aircraft raids (out of a total of 241 during the day), those at Brie and Béthencourt being each attacked four times. In addition to these attacks by bombers about ninety individual attacks with light bombs were carried out from a low altitude against troops and transport crossing the Somme bridges, by fighters of the 9th Brigade who were engaged on offensive patrol and escort duty over that area. Bombing attacks during this period were also directed against railway objectives at Chaulnes, Harbonnières, and Étricourt, and by No. 18 Squadron of the 1st Brigade to the north against Cantin, on the Douai–Cambrai line. Large troop movements in Bray and Fouceaucourt were also bombed, but it will be noted that about 70 per cent. of the bombing effort in the first twenty-four hours was on the Somme bridges.

Throughout the day No. 48 (fighter-reconnaissance) squadron had single aircraft watching for road and rail movement in the army reconnaissance area, which apparently extended at its deepest point about twenty miles over to Péronne, and laterally from Nesle in the south to Bapaume, a distance of some twenty-five miles. The information received from these reconnaissances was disappointingly meagre—perhaps not unnaturally so in the light of what is to-day considered a fair reconnaissance area for a single pilot.

A boundary between the British air force and the aviation of the French First Army had been agreed upon, and south of the railway Amiens–Roye the French air units operated on similar lines, though unfortunately few details can be discovered of their activities during this battle.

August 9th. Orders were issued on the evening of the 8th for the advance to be continued next day to the line Roye–Chaulnes–Bray-sur-Somme. The attack on the 9th did not begin until about 10 a.m., and the resistance encountered varied largely in determination on different parts of the front. No serious counter-attacks were made, but the enemy did succeed in 'puttying up' the gaps in his line by reinforcements brought in by march route and motor transport. As the day progressed the German resistance was felt to be stiffening, but the net

'result of the day's fighting was another big advance on the whole Army front, extending to as much as 9,000 yards in the South.

The line we had now reached ran approximately Bouchoir–Rouvroy-en-Santerre–Méharicourt–Framerville–Méricourt (excl.) . . .–Dernancourt (several miles short of the general objective allotted). On the right of the Fourth Army the First French Army had also made progress and reached the general line Pierrepont–Arvillers. . . . It was still felt that if the determined pressure exerted on August 8th and 9th was continued the enemy's resistance might be definitely broken down.'[1]

In the air the low-flying fighters continued to operate as on the previous day, with the exception that two squadrons, Nos. 23 and 48, were returned to the upper air and employed entirely on offensive patrols, less one flight of 48 which continued to do the medium reconnaissance in the army area, with much the same results as on the 8th. This left only two fighter squadrons on the front of the Australian Corps, since one of the squadrons previously allotted to that Corps sector—No. 41—was transferred to that of the Canadian Corps which was taking the principal part in the day's attack. As on the 8th the low-flying squadrons spent the day attacking troops and transport in the area west of the Somme.

It was not only on the ground that the enemy's resistance began to stiffen on the 9th of August. It soon became apparent that the enemy had concentrated air units from other sectors to meet the attack on the Somme; a largely increased number of enemy aircraft were encountered and heavy air fighting took place. The fighters of the 9th and 3rd Brigades continued their duty of offensive patrol and indirect protection of the army co-operation and low-flying aircraft, reinforced by the 5th Brigade squadrons already mentioned.

Two additional night-bomber squadrons, Nos. 207 and 215 of the 9th Brigade, were added during the second twenty-four hours of the battle to the eleven bomber squadrons already engaged. The important railway centres of Douai and Cambrai were attacked, and the detraining station at Péronne. But, as on the 8th, the great majority of the bombing raids were again directed against the Somme bridges, one of which, that at Brie on the main Amiens–St. Quentin road, was attacked no less than eight times by bomber formations during the day. Altogether during the 9th of August 173 aircraft attacks were

[1] *The Story of the Fourth Army*, p. 57.

made on the Somme bridges as against thirty-three on other targets.

August 10th and 11th. The objective laid down in orders for the advance on the 10th was the same as that previously ordered for the 9th. But the enemy had succeeded in stabilizing his defence, and in spite of desperate fighting, in which some ground was gained, little progress was made in the face of determined counter-attacks by the Germans. Nevertheless orders were issued on the evening of the 10th for the attack to be continued on the 11th with a view to pressing the enemy back to the Somme and establishing bridge-heads on the right bank of that river. But on the morning of the 11th the attack was called off by the Canadian Corps Commander in view of the difficult ground and the lack of tank or adequate artillery support. An attack by the 32nd Division on the southern end of the Army front was launched before it could be countermanded, but was checked by machine-gun fire; and although the Australians captured Lihons and were able to hold it against very heavy counter-attacks, by the afternoon of August 11th the impetus of the Fourth Army's attack was finally spent. A conference was held at Villers Bretonneux at 15.00 hours that afternoon: 'from all the reports that had been received it was quite evident that the enemy's resistance had stiffened and that he had been able to bring up fresh troops and to reinforce his shattered artillery'.[1] The army commander had no intention of persisting in attempts to break through regardless of cost, and he decided therefore not to press the attack but to reorganize and bring up the artillery with a view to a fresh attack at an early date.

In the air, low-flying attacks continued on the 10th in the same area as before; but the increased activity of the enemy in low-flying attack and against our aircraft engaged in this duty is reflected in the orders of the 22nd Wing for the 10th, in which patrols of four aircraft without bombs—drawn from the low-flying squadrons of the previous day—were detailed throughout the morning for the purpose of destroying enemy low-flying aircraft and protecting our own. Low-flying attack was not ordered for the 11th. The efforts of the 22nd Wing were allowed to slack off—which was probably very necessary in view of the strenuous work and heavy casualties of the three

[1] Ibid., p. 63.

preceding days. The fighters returned to their more accustomed duty of offensive patrol—with the specific task again of destroying enemy low-flying aircraft and protecting our own, though their orders were, as those of the fighters engaged on similar tasks had been on the 10th, to come down and shoot up ground targets if available towards the end of their patrols.

All thirteen bomber squadrons were again engaged on the 10th, though on the 11th the night-bomber squadron, No. 101 of the 5th Brigade, does not appear to have been in action. The proportion of bombing effort allotted to the Somme bridges and other targets began to change; 105 attacks being directed against the bridges on the 10th out of a total of 198, whereas on the 11th 118 attacks were made against other targets as against only 25 on the Somme bridges. Péronne in particular received 10 raids on the 10th, and in the last two days of the battle the bombing effort shifted from the Somme crossings mainly to objectives on the railways, including Cambrai, Douai, Somain, Roisel, and Équancourt, the latter being attacked seven times during the two days.

During the lull in the battle after the cessation of attacks on the 11th, orders were issued that the bombing of the Somme bridges and of rest billets, roads, and railways was to be continued by day and night.

The German Air Forces under the orders of their Second Army totalled 134 aircraft, of which 64 were single-seater fighter and 'battle' (or low-flying attack) aircraft and 12 were bombers. And although the air units of the Seventeenth and Eighteenth Armies on the flanks, which were very approximately of the same strength, were also available to take their part in the defence, still the German air concentration at the outset—owing to the complete unexpectedness of our attack—did not represent anything approaching the proportion of their total strength which they had concentrated for the battles earlier in the year, when they were on the offensive, and was numerically very inferior to our own. It is worth noting that on the whole British front the serviceable bombers and fighters on August 8th outnumbered by about four times the German aircraft opposed to them (the actual figures being 1,390 to about 340).[1] And

[1] The British figures are correct; the German approximately so.

whereas our excellent system of provision and supply enabled us to replace our casualties with the minimum of delay, it was often a matter of weeks before the enemy could replace his lost or damaged aircraft. This being so, the German air service deserves the greatest credit for having given us the fight that they did, and it is hardly surprising that their tactics inclined to the defensive, rather than the offensive.

On the day of the attack no very serious air opposition was encountered, and the majority of our casualties—which were heavy—were among the low-flying fighters, and were due to small-arm fire from the ground. Following the attack on the 8th came two days of the most intense enemy activity on the Fourth Army front, 305 aircraft being observed on the 9th and 263 on the 10th. This was, of course, to be expected, the next heaviest days being the 22nd and 23rd when 180 and 181 respectively were observed, this slight increase coinciding with the second phase of the battle. The effect of this increase of enemy air activity in diverting a proportion of our fighters from low-flying attack to offensive patrols and protective duty has already been noted; and a further result was that on the 10th our bombers attacking the Somme bridges were forced up to heights of 10,000 feet and above, from which bombardment of such small targets ceased to be very effective. Apart from these four days above mentioned enemy air activity was noticeably restricted, an indication that at this period of the War the Germans were obliged to husband both aircraft and pilots. During the month of August, however, the enemy air forces on the Fourth Army front were reinforced by some of his best units, amongst them the Richthofen fighting squadron;[1] and on several occasions for limited periods as many as fifty enemy aircraft could be seen in the air at once. These periods usually coincided with further retirements on the part of the Germans, whose object appeared to be mainly to prevent our reconnaissance aircraft getting across and observing their movements.

From the 9th of August for the rest of the month there was a marked increase in enemy bombing at night. To counter this searchlights were advanced and grouped well forward, a continuous belt of lights being formed across the army front,

[1] Richthofen himself was dead, having been shot down on this front about four months before.

the most forward being in action from 5,000 to 8,000 yards from the line.   And it is a tribute to the efficiency of our A.A. defences—as well as a measure of the immense resources then available—that from the time this belt was formed bombing of back areas became almost negligible; in fact by September bombing was almost entirely confined to the 'forward areas', in other words to the zone which was within enemy gun range.

Low-flying attack against troops and transport was not confined to ourselves.   The Germans had units specially trained for this work, known as battle-flights, and many effective attacks were launched which imposed some delays on our advance on the 9th and 10th; a notable instance being the advance of the 10th Australian Brigade along the road Amiens–Brie on the evening of the 10th, which was stopped by intensive low bombing and artillery fire.

The success of the British attack on August 8th, 'the black day of the German army', came as a moral tonic of inestimable value to the Allied forces after the long, depressing months of retreat and defensive fighting.   And as we now know it had an immense psychological effect on the minds of the enemy commanders, and led both the Kaiser and Ludendorff to conclude that victory was impossible, and that the War must be brought to an end.   But this need not blind us to the fact that, judged by the standard of what it was intended to achieve, the immediate results of the battle of Amiens were disappointing. 'Its initial penetration of six to eight miles, and ultimate twelve miles were again excellent by 1915–17 standards, but in March the Germans had penetrated thirty-eight miles in the reverse direction without achieving any decisive result.'[1]   Roye and Chaulnes—on which the advance was to be pressed 'with the least possible delay' after the capture of the Hangest–Harbonnières line—did not fall until three weeks later in a subsequent attack, and Ham not for a month.   The tone of confident elation on the first evening, after the overwhelming success of the break-in on the 8th—'There were few hostile reserves immediately available and the prospects of further success on the following day were extremely bright'—gives way to one of more measured optimism on the evening of the 9th, 'It

[1] Liddell Hart, *The Real War*, p. 457.

was still felt that if the determined pressure exerted on August 8th and 9th was continued, the enemy's resistance might be definitely broken down.' And finally, thirty-six hours later, the enemy's resistance has stiffened, and there is no intention to try to break through regardless of loss. The 8th of August, in short, was a completely successful break-in; it failed to develop into a break-through. Various reasons have been adduced for this failure to fulfil the radiant promise of the first day: lack of reserves, the difficulty of the ground on the old Somme battle-fields, the difficulty of getting forward sufficient supporting artillery, and the exhaustion of the troops—particularly of the tanks. All these undoubtedly contributed to the result, but above all was the fact (to which insufficient importance is normally attributed) that the enemy 'had been able to bring up fresh troops and to reinforce his shattered artillery'. Actually, by the evening of the 11th, sixteen German divisions had been identified by prisoners on the Fourth Army front over and above the eight which were holding the line on which the blow fell; and of these, twelve were brought in from other German army fronts, some of them from as far away as Ghent and Courtrai in the north. In this fact lies the principal interest and perhaps the most important lesson of the battle of Amiens from an air point of view. Even to-day the capacity of air action to hinder or even totally prevent the movement of enemy reserves is not sufficiently appreciated. So there is little wonder that it was not realized in the first war in which aircraft played a serious part. As at Amiens in August 1918, so at Cambrai in the previous November, during the nine days after zero a hundred trains a day poured into the threatened front thirteen reserve divisions and several hundred additional units of various kinds. And the great German breaks-in in the spring of 1918 were all brought to a standstill, owing partly to the exhaustion of the attackers, but mainly to the fact that the railways and motor-lorries of the defence were always able to rush reserves to the threatened point in time.

In the event, the R.A.F. made two contributions of great importance to the success of the initial attack on August 8th. First, the complete surprise achieved was largely due to the high degree of air superiority prevailing—which, indeed, could hardly have been otherwise in view of the fact that the British aircraft outnumbered the Germans between the Somme and the sea by over three to one. Secondly, the action of the low-flying fighters on the 8th was a factor of first-class importance in the overwhelming success of the initial break-in. But apart from these two factors it is impossible to assert with any confidence that the result of the battle after about 14.00 hours on the 8th would have been materially different, or that the ultimate line reached and held by our forward troops on the 11th would have been materially short of where in fact it was, if not a bomb had been dropped or a round fired by aircraft against ground objectives.

If this be so it is a damaging admission, in view of the fact that this battle saw the greatest concentration of air strength of any battle of the War; and it is the object of the following chapters to examine why it was so, to suggest certain directions in which—by the light of our present experience and knowledge —the situation might have been altered, and to draw certain deductions for our guidance in the future.

IX

THE R.A.F. IN THE BATTLE

THE AIR PLAN

'In every operation of war it is essential to decide upon and clearly to define the object which the use of force is intended to attain.'[1]

THE system whereby the arrangements for air co-operation in battle during the last War were so often made by means of personal discussion between the Staffs, unconfirmed in writing, has already been referred to in the introduction to this book. And a study of the air plan for the battle of Amiens seems to emphasize the dangers of that system. The Commander and Staff of the Fourth Army were fully alive to the value and importance of air action; the Air Officer Commanding the 5th Brigade, R.A.F., was constantly called to Army Head-quarters, and took part in long and detailed discussions on the employment of his squadrons in the forthcoming operations. But, apart from the instructions summarized on a previous page, the army battle orders contained no reference to the action of the air force. And in particular the *object* of the air operations— the effect they were intended to produce, the part they were to play in the Plan as a whole—was not clearly defined on paper. This may have been explained to General Charlton verbally by the Staff, but the fact remains that he evidently did not understand it—or understood only part of it. It seems not impossible that the development of the plan, described in the last chapter, from a limited operation to secure the old outer Amiens defences, was not made clear to him. For in his instructions which were read out to every pilot and observer throughout the 5th Brigade on the afternoon of August 7th he described the task of the infantry as being only the capture of the blue line—and that of the other arms, artillery, cavalry, and tanks, to support the infantry on to the blue line. And with this in view he defined the task of the air force as being of wider scope, namely, not only the direct support of the infantry on to the blue line, but also to help the other arms to help the infantry. From these instructions—slightly reminiscent of the House that

[1] F.S.R. ii, sect. 7. 2.

Jack built—the object laid down for the air force emerges as, directly or indirectly, to *help the infantry on to the blue line.*

It is, perhaps, permissible to reflect that this was not a very wide scope for an arm of which the rate of tactical movement was even then at least twenty times that of the other arms on the ground. It is, of course, important to try to place oneself in the position of the commanders at the time; and to realize that in their view—and probably in fact—success in this attack, or in other words the capture of that blue line, meant the difference between winning the War and—at best—a stalemate and a negotiated peace. This being so, the orders issued from Army Head-quarters and the instructions by the Commander of the 5th Brigade did in fact effectively secure the attainment of this rather limited object. All units thoroughly understood the situation and the part they were to play in it, and carried out their orders with complete success, and ·with the utmost self-sacrifice and gallantry. The mistake lay in the selection of too limited an object, and in a failure to look sufficiently far ahead. The minds of men attuned to years of static warfare, accustomed to the desperately contested trench-to-trench offensives of the Somme, the Ypres Salient, and Arras, and fresh from the months of defensive fighting of the spring and summer of 1918, perhaps found it difficult thoroughly to grasp the scope of this operation as a definite break-through—the first act in a drama of open warfare once again. For in the strategical plan outlined by the Supreme Command—the advance with all possible speed on Roye and Chaulnes and subsequently on Ham—the capture of the blue line, though an essential and by far the most important and difficult stage, was nevertheless only the first stage. No longer—as in the original instructions—was the blue line when captured to be elaborately organized as a main battle position, but it was to be the first bound in a far deeper penetration. In the difference between this idea and the idea which emerges from a study of the instructions probably lies the explanation of much that happened in the subsequent days, and of much that there is to criticize in the direction of the air effort in this battle. Charlton told his officers that the battle was intended to be a one-day battle, and that the blue line—the old Amiens defence line—was to be the ultimate objective. And this suggests that in spite of the excellent liaison between the A.O.C.

and Army Head-quarters, the absence of any clearly defined object in orders is not altogether explained by a lack of written confirmation of verbal instructions, nor compensated for by the system of personal discussions. There never had been a one-day battle in this War, nor should our knowledge of German methods or our experiences at Cambrai in the preceding December have led us to suppose that there ever would be. Even if the blue line had been our ultimate objective, its capture on the one day was morally certain to be followed by counter-attack on the next. But the blue line was not our ultimate objective, and was in fact mostly in our hands early in the afternoon of the 8th. Although in issuing orders to land forces there is reason for not legislating for more than a very limited period ahead, these reasons do not apply in the air, and our instructions to the air force should have provided not only for the initial capture of the blue line, but further for the action with a really wider scope against the enemy reserves and re-inforcements that were bound to be rushed in to stem the subsequent advance.

So in this battle—as, indeed, in any battle on land—the *object* of the air force, or perhaps it would be more accurate to say the intention of the commander for the air force, should be to *isolate the battle-field from enemy reinforcement and supply*—it being, of course, understood that at the same time and largely by the same means our own lines of communication must be protected against enemy air action. A paragraph to this effect should be included in the instructions issued to the A.O.C.; and these should also include an indication of the proportion of available effort intended by the Commander-in-Chief to be directed against any particular objective, or rather class of objectives, such as rail communications likely to be used by distant re-inforcements or the movement of the enemy's immediate reserves and supply traffic on the roads. Consequent upon this, the Plan would be worked out in detail and targets selected by the Air and General Staffs, in consultation with their Intelligence branches, and assisted by the technical advice of the transportation and supply authorities; and finally the A.O.C. would issue his orders to the squadrons as to the method by which the plan should be put into execution. Supposing the object had been decided upon on these lines on August 8th, and every available resource directed to its attainment; supposing the A.O.C. had succeeded in blocking the main approaches to the area Bapaume–Le Catelet–Guise–La Fère–Noyon, and checking the westward movement of enemy columns within that area: is it unreasonable to suggest that the impetus of our initial advance might have been maintained unchecked across the old Somme battle-fields and on to St. Quentin, finally rolling up the German flank on the Aisne? An optimistic estimate, perhaps, and one which the experience then available probably did not entitle us to form. But such might have been the object, and the following pages may give some idea of what chance it had of being successfully achieved.[1]

This, then, should be the governing principle in the employment of the air striking force in a battle on land—isolation of the battle-field from enemy reinforcement and supply. And it need not detract from the value of this principle that on August 8th it was necessary, and may again be necessary, to divert a proportion of the striking force temporarily to another role. It has been explained that, in the view of those responsible at the time, the capture of that blue line very likely meant the difference between winning the War and losing it; and it has been suggested in an earlier chapter that when the problem is to *break the crust* of highly organized defences, then assault aircraft may have to be diverted in the initial stages to the close support of the attacking troops. So on August 8th it was certainly sound—and in any case psychologically inevitable—that some of the fighters should be employed on the first day this side of the blue line, dealing with the immediate opposition in the enemy forward areas and particularly with anti-tank guns. And the most effective direction of the efforts of the fighters so employed might have been secured by placing them temporarily under the command of Corps, as suggested in an earlier chapter. Corps Intelligence Staffs alone probably had the sufficiently detailed information from air photographs and other sources of the enemy defensive system; and moreover Corps will usually

---

[1] It may be argued that such an ambitious plan could not have been carried through with the reserves immediately available behind the Fourth Army on August 8th. This is obviously true; but it is fair to assume that had the Supreme Command formed the opinion that the correct application of air power made such an object possible, they could have taken the necessary steps to ensure that sufficient reserve divisions were available to carry it to a successful conclusion.

be the highest formation in a position to get information, *in time* for the assault aircraft to act upon it, of any situation which requires their immediate attention, such as that which developed at Flesquières during the battle of Cambrai quoted in a previous chapter.

THE PLAN AS IT WAS

It is clear from what actually happened on the 8th that the object laid down in the 5th Brigade instructions was not in fact interpreted so narrowly as to mean that the whole effort of the air striking force was to be expended in direct close support of the infantry on the battle-field itself. The squadrons allotted to this role were *the fighters* of the 22nd Wing, nine squadrons (less one flight), a total of 176 aircraft; their orders were to carry out low-flying attacks on hostile troops, transport, gun-teams, ammunition wagons, balloons on the ground, and last but not least anti-tank guns. Zones in which the squadrons were to operate were detailed, aircraft were to carry 20-lb. bombs and a full load of ammunition, and were to leave the ground in pairs from each squadron at intervals of half an hour from $Z+20$ minutes. These orders were admirably carried out in spite of severe casualties; and there is no question but that the action of the low-flying fighters was a factor of immense importance in the overwhelming success of the initial attack, and put the final seal on the fate of the blue line. Their subsequent employment *after* the capture of the blue line is more open to question, and there can be little doubt that they would have had far more influence on the eventual outcome of the battle had they been used from the 9th onwards to attack columns on the march in the area east and north of the Somme.

On the other hand the tasks allotted to the *bombers* for the first twenty-four hours of the battle were in fact evidently designed to interrupt the arrival of enemy reinforcements in the battle area. During this period eleven day and night bombing squadrons, a total of 201 aircraft, were engaged against objectives in connexion with the battle. The proposals for their employment submitted to G.H.Q. by the Air Officer Commanding the R.A.F. in the field were based on the consideration —amongst others—that it was no use bombing railway stations until the evening of the attack, as there was likely to be little activity for the first ten hours; and that distant railway stations should not be bombed till after the first twenty-four hours, when troops might be coming in from a distance to make a counter-attack on a large scale. Thus the co-ordinated programme for the employment of the R.A.F., issued by G.H.Q.,[1] provided on Z day for 'Bombing attack on enemy aerodromes in early morning. Fighter squadrons to work in conjunction with bombers, and subsequently to be held in readiness to counter enemy activity on Fourth Army front. Bombing attack on enemy communications in the evening, Fighter squadrons to work in conjunction with the bombers.' During the night Z/Z+1 'Two night-flying squadrons (one Handley-Page) to continue the attack on hostile communications carried out by day-bombers in the evening. A second Handley-Page squadron to attack communications leading to Fourth Army front from the North.'

This was the air plan in outline. Unfortunately there is no record of the process by which this plan was evolved. No doubt the details were carefully considered and discussed by the A.O.C. 5th Brigade with the Staff of the Fourth Army, and by the Air Staff at R.A.F. Head-quarters with the General Staff at G.H.Q. There is, however, no record of anything in the way of an appreciation—even of the limited type produced for the Passchendaele battle, referred to in a previous chapter. Some reasoned appreciation there must have been, but it must be admitted that its results are not apparent in the direction of the air striking force during the battle. There seems, indeed, to have been a singular neglect in this air plan of the three essentials described on pp. 88-90—Information, Expert advice, and Reconnaissance. This was not because these things were not available. As for *Information*, the Intelligence Staff had at their disposal a vast mass of the most detailed and circumstantial information about the enemy. Their conclusions may not always have been sound, but their information on points of fact was comprehensive and usually accurate. Order of Battle maps showed every enemy division on the Western front, their positions if they were in the line, their billeting areas if they were in rest or reserve, and their approximate fitness for battle. More detailed intelligence maps and files showed the positions

[1] See Appendix A.

of head-quarters, billeting villages, railheads, detraining stations, hospitals, camps, and dumps. To what extent all this
mass of information was used in framing the air plan is not
known; but from the measures attempted, and from the results
achieved, one gains the impression that full advantage cannot
have been taken of it. We know that the Intelligence Staff
anticipated that the Germans would be able to reinforce the
front attacked to the extent of eight divisions from reserve by
the evening of August 11th—an estimate which fell short of
what actually occurred by about 100 per cent. But there is no
evidence to show how they arrived at that figure, or whether
they told the air force which divisions they thought could be
brought in, from where or by what routes.

The use of *Technical advice* about the vulnerable points in
the enemy system of communications can hardly have arisen,
because there was no really serious attempt to cut the enemy
railway communications on a carefully thought-out plan. That
this point was not altogether ignored is evident from the G.H.Q.
order (Appendix A) and from the recommendations of the Air
Staff outlined on p. 169. The latter proposals were in accordance with the policy—originally laid down in February 1916[1]—
of confining attacks on railway communications to periods when
active operations were in progress. But it is arguable that this
policy was carried too far in this battle. The point made by
the Air Staff to G.H.Q. that no important rail movements
would be taking place until the evening of the 8th at the earliest
was no doubt true. Nevertheless it was as safe as anything can
be in war to assume that within the first twenty-four hours there
would be heavy train movement, for instance through Cambrai
and probably also Bohain. Even before train movement on a
large scale began, a series of attacks directed against these
places with the intention of breaking the permanent way,
destroying the signal and watering gear, and blocking the lines
with debris would surely have resulted at least in very serious
disorganization and delay in the enemy's arrangements for
strategical mobility. As it was, Cambrai was not attacked till
the night of the 9th, by which time at least one division (and
probably more) had passed through the junction unchecked.
So that—although during the first ten hours of the battle as

[1] See p. 126 above.

suggested by Air Head-quarters there were more urgent and profitable objectives for the bombers elsewhere than on the railways—by the evening of the 8th the main weight of their attacks might more profitably have been switched from closer objectives to the main railway approaches, and particularly to Cambrai. Actually on the morning of the 8th the four aerodromes already mentioned were duly attacked at dawn, but after that, for the three days 8th to 10th inclusive, the greater part of the bombers' effort was employed in an attempt to put down a sort of barrage of bombs on the Somme crossings;[1] beginning on the 8th with twelve different bridges on a thirty-mile stretch from Bray to Pithon, and gradually contracting to the four most important objectives from Péronne to Falvy on the 10th. The only objectives farther back on the enemy communications raided in the first twenty-four hours were Cantin, on the line Douai–Cambrai (5 D.B.), Étricourt station (10 D.B.), the road Amiens–St. Quentin at Fouceaucourt, west of the Somme (10 D.B.), and railway objectives at Harbonnières (13 D.B.) and Chaulnes (7 D.B.). On the 9th seven Handley-Pages were sent to bomb Cambrai station (7 N.B.), and the stations at Douai (7 D.B.) and Péronne (10 D.B.) were attacked. On the following day, the 10th, the weight of attack began to shift from the Somme bridges to objectives on communications farther back; and during these twenty-four hours attacks were carried out against Péronne station (27 D.B.), Équancourt station (11 D.B., 18 N.B.), the stations at Cambrai (7 N.B. Handley-Pages), Douai (6 D.B.), and Denain (9 D.B.), and the road centre in Bapaume (10 D.B.). But on the first three days after zero the proportion of bombing effort, reckoned by individual aircraft attacks, directed against the Somme bridges and other objectives was on the 8th 162 to 79, on the 9th 173 to 33, on the 10th 105 to 93, until finally on the 11th the greater part of a diminished effort was directed against railway communications, 118 out of 143 attacks being carried out against this class of objective. Unfortunately, by this time twelve enemy reserve divisions from other army fronts had succeeded in reaching the battle-field in support of the shattered Second Army.

From a glance at the map the idea of a sort of barrage on the Somme bridges looks, at first sight, attractive. But from Bray

[1] See Appendix B.

to Pithon there are fourteen permanent road and rail bridges, apart from a number of foot bridges and military bridges which existed at the time; and by the time they reached the Somme enemy columns could afford to split into many smaller groups, and in effect the targets presented, instead of being like a hose-pipe, became like the nozzle or spray at the end of it, and were correspondingly far more widely scattered and less vulnerable. Moreover, the Somme crossings were only a few hours' march from the front; and though undoubtedly the Germans suffered severely under the bombing, there were constant long pauses during which no bombing took place, and the enemy columns could—and in fact did—pour across the bridges and on to the front line. Very much the same thing applies to the detraining stations such as Péronne, Nesle, and Équancourt; it is true that troops detraining are very vulnerable, but they are equally vulnerable when entraining. In summary it amounts to this, that, if the aim is to prevent the flow reaching its destination, it is better to block it at the cistern or the tap or to cut the hose-pipe than to try to block up all the holes in the spray.

Finally, it has already been suggested in a previous chapter that bridges are not as a rule suitable bombing objectives. These small bridges over the Somme presented very small targets; they were mostly solidly built and almost impossible seriously to damage—certainly with the bombs then available; and moreover even if broken they are not difficult to repair— for example, an eighteen-foot breach in the bridge over the Luce at Hangard was made passable in a few hours by Canadian Engineers on August 8th. It would have been better to regard them merely as defiles, to attack the troops and transport that had to cross them; and this was a task for which, in the circumstances at the time, assault aircraft would have been better suited than bombers flying at the heights which the existing conditions evidently made necessary,[1] especially later on in the four days' battle.

The third essential in any sound air plan—adequate and properly directed *Reconnaissance*—was notably honoured in the breach rather than the observance during the battle of Amiens. We know what happened: this was an operation of absolutely first-class strategical importance; the Intelligence Staff had

[1] See p. 161.

anticipated that eight German divisions could reinforce the front within the first four days—and actually we know that double this number arrived; it was known that the bulk of these divisions must be brought in from other Army fronts, and in fact bombers were detailed to attack communications leading from the north; and finally we had concentrated a very strong force of bombers of which the primary task was attack on communications.   Yet of the three squadrons of first-class fighter reconnaissance aircraft which were engaged in the battle of Amiens, two were used solely as fighters, on offensive patrols and as escorts to bombers; while of the third (No. 48 squadron, under the direct orders of the Fourth Army) two flights were used for low-flying attack, leaving only *one flight* of seven aircraft to meet the reconnaissance requirements of this vitally important situation.[1] How was this flight employed?   The officer commanding No. 48 squadron was given orders to carry out reconnaissance of the Army area (which had been described in a previous order as from Aveluy to Moreuil—presumably points on the lateral boundaries with the reconnaissance areas of the Third British and First French Armies) at close intervals throughout the day; and was told that his reconnaissance aircraft should not fly too high; no doubt these orders were supplemented by more detailed instructions through the Intelligence liaison officer of the Wing.   But what actually happened was that reconnaissances were done by single aircraft at intervals throughout the day manned by flying-officers, and in one case even by a non-commissioned officer, as observers.   Eight reconnaissances were done on August 8th, of which one included Péronne and all the rest were over the area well this side of the Somme; of the four reconnaissances on the 9th, again only one went as far east as Péronne.   On the 10th only one reconnaissance appears to have been done, which went as far as Tincourt (5 miles NE. of Péronne).   But on the 11th there was more activity, and seven patrols in the course of the day watched the area Bapaume–Péronne–Ham–Nesle–Roye–Rosières, and reported considerable railway activity.

But the activities of these gallant junior officers and non-

---

[1] Apart, of course, from close reconnaissance—contact and counter-attack patrol —by the army co-operation squadrons on the actual battle-field, which does not affect this point.

commissioned officers in the course of these twenty patrols were almost entirely wasted. They produced little information which could not perfectly safely be taken for granted, or which in fact had any effect whatever on the course of the battle. For instance, once the initial break-in had succeeded as it did, it would have been perfectly safe to assume that by 13.00 hours there would be considerable movement of transport eastwards from the vicinity of the blue line. It could equally have been taken for granted that on the 10th and 11th there would be a certain amount of rail movement in the vicinity of such stations as Péronne, Nesle, and Ham. What *was* important to make sure was where those trains were coming *from*, and what movements there were on the main roads and railways, not immediately behind the front, but away back at least as far as Douai, Denain, and Cambrai to the north, and Laon and La Fère in the south. These questions were left unanswered—*because they were never asked*. Further, although it must be assumed that some information must have been brought back by bombers and fighters who had been attacking the Somme bridges, no reconnaissance was dispatched with the special task of reporting as to which of these bridges were being principally used, so that the bombing could be concentrated where it would have most effect.

To sum up. A sound initial appreciation must be followed by well-directed reconnaissance to confirm or modify the conclusions arrived at. Bombers should not be sent out blind to attack unreconnoitred objectives which may have little or no effect on the course of the battle. And finally we must make full use of the mobility of aircraft to send our reconnaissance *deep*, to give the bombers ample warning and room in which to stop the flow of hostile reinforcement and supply.

THE PLAN AS IT MIGHT HAVE BEEN

In considering the process by which the object suggested on p. 167 might be translated into a definite plan, it is necessary to bear in mind a condition which anticipates conclusions arrived at in a later chapter. It must be remembered that this battle was to be fought not only by the British Fourth Army, but also by the French First Army on its right, and therefore from the air point of view the employment of the British and

French air forces must be considered as one problem. Unfortunately this is rendered more difficult to-day by the lack of information about the air forces available with the French First Army or how they were used. But although for this reason detailed consideration can only be given to the British share in the battle, it must not be forgotten that the air operations were —or should have been—those of an inter-Allied air force against the German land and air forces opposed to the Allies on the whole battle-front from Albert to Montdidier.

AN APPRECIATION

The extent and scope of the *information* at the disposal of the commanders before the battle has already been described. Based on this information and a study on the map of the enemy's system of communications, the first step is to try to estimate the enemy's probable courses of action and the *forces available* for the reinforcement of his Second Army front. Forward of all there were, of course, the troops in support, the reserve battalions of regiments and the reserve regiments of divisions actually holding the line, the approximate locations of which could be discovered from the detailed intelligence maps available; these and the enemy battery areas would have to be the first objectives of any squadrons which might be detailed temporarily to the direct support of the infantry as far as the blue line.

Next to be considered were the divisions in reserve actually in the German Second Army area,[1] from which it might be assumed would be drawn the first reserves that would be thrown in, and of which there were six. In immediate reserve west of the Somme were the 43rd Reserve and 109th which were known to be at Flaucourt and Berny. East of that river were the 54th at Roisel and the 243rd at Ribecourt (south-west of Cambrai), both of which were fit for battle, having been out of the line for a month, and could therefore be counted on with certainty as early reinforcements. East of Le Catelet was the 107th division, which had only been taken out of the line in the first week of August, and the 21st, which had also fairly recently been relieved and was therefore counted as a tired division.

Next, it might be assumed that we should have to reckon with

---

[1] See Sketch-map no. 2 at the end of this volume.

the fit[1] divisions in reserve in the army areas on the immediate
flanks of the Second Army. To the north, in the Seventeenth
Army area, were known to be the 185th at Dury, north-west of
Cambrai; the 5th Bavarian at Marquion; the 5th Bavarian
Reserve at Cambrai; the 26th Wurtemburg Reserve south of
Bapaume, and the 221st north-west of Denain. In the Eighteenth Army area on the other flank were the 82nd Reserve,
north-west of Guiscard, the 84th Reserve and the 5th Reserve
divisions at Esméry-Hallon and Roye respectively.

Finally, there were the divisions in reserve behind other
sectors of the Western front, still farther away from the battle
area. These were, of course, much more a matter for specula-
tion; and probably the most that could be said was that almost
certainly the bulk of the reinforcements drawn from more
distant sectors would come from the north. South of the Eighteenth Army the enemy was known to have comparatively few
fit divisions, even in the line—actually the Seventh Army comprised no less that thirty-five unfit divisions, the result of Pétain's
counter-attack on the Marne in July. But in the north,
especially in the Sixth Army behind the Lys Salient, were many
fit divisions which had been assembled in preparation for
Ludendorff's next offensive in that sector.

So much for the probable reinforcements and their billeting areas before the battle. Now to consider *the most probable
routes* which they would use to reach the threatened front. An
estimate of this factor must be based on a study of the map and,
of course, to some extent upon conjecture; and it must be confirmed and amplified, as soon as the attack begins and surprise
is no longer a consideration, by immediate and unremitting
reconnaissance. But some estimate is obviously necessary to
enable us not only to make our initial attacks where they have
most chance of being effective, but also to direct our reconnaissance intelligently.

It will be remembered that the object in view as far as the
air force is concerned is to isolate the battle-field from enemy
reinforcement and supply. At this stage this object needs some
further definition. To the arguments already quoted against
the plan of a bomb barrage on the Somme bridges there must

---

[1] By a 'fit' division is meant one that was known to have been in rest or reserve
for a month or more.

be added the consideration that even were it successful in preventing the enemy reinforcing the area south and west of the
Somme, it would still have enabled him to concentrate in force
to oppose our advance east of that river.   A reasonable basis
for the plan must therefore be on a wider scale, and might be—
as already suggested—to deny to the enemy the area between
the River Oise and a line drawn from Bapaume to Guise; that
is to say, to prevent reinforcements entering that area and to
break up the movement towards the battle-front of enemy
reserves already within it.   If this could be successfully achieved
it would enable the attack to maintain its initial impetus unchecked across the Somme to the great road and rail centre of
St. Quentin and beyond.

First as regards *movement by road* into this area:

The map shows that the country east of the line Noyon–
Bapaume is very well served by roads, and therefore any attempt
to block all the road approaches to the battle area would be open
to an even greater extent to one of the objections against the
Somme crossings plan, namely that far too many objectives
would be involved.   Moreover, to block or breach a road by
bombing is far more difficult than to do the same to a railway,
and save in exceptional circumstances is not in fact a practical
operation of war.   There are, however, a number of defiles,
where roads converge in villages and towns such as Bapaume,
Fins, Le Catelet, La Fère, and Noyon, which are eminently
suitable for bombing attack *when columns are actually passing
through them*; and there are a number of long stretches of road
with little or no cover, where low-flying attack is likely to be
extremely effective.   Nevertheless, with the exception of certain
formations in immediate reserve, attack on reinforcements
moving by road should mostly be as the result of information
obtained by air reconnaissance of large moves actually in
progress.

By railways, on the other hand, the area is not so well served.
Breaches at only five points, at Cambrai, Le Cateau, Le Nouvion,
Vervins, and Laon, would virtually put a stop to all rail movement towards the front attacked.   Outside the semicircle so
described the railway system is considerably more elaborate,
particularly to the north and north-east towards the industrial
areas on the Belgian border whence the bulk of distant re-

inforcements were expected to come, and it would be necessary to cut lines at more than twice that number of points. But although the places named in the north are only two days' march from the front, still the delay of two days, added to that resulting from air action against the marching columns when they had detrained, would probably be fatal to the enemy. Therefore, in view of the essential importance of intensive and sustained attack, the five points named above should be those selected at which definite blocks are to be attempted; although, especially at night, trains on the move might be attacked and derailed on either side of the points named.

It will be noted that the interruption of reinforcements only has been mentioned, and not the dislocation of supply. It is true that if the German Second Army could not be reinforced, the fact that it could not be supplied would be comparatively immaterial. But neither is there any doubt that, if the programme of attack on communications outlined above were successfully carried through, the supply situation both as regards ammunition and other supplies would be desperate.[1]

The temporary and subsidiary task of direct support for the infantry and tanks on to the blue line needs no comment; in actual fact the tactics adopted worked admirably, and the assault aircraft inflicted great loss and demoralization upon the enemy's forward divisions.

Based on the consideration of these factors and a number of others (including the strength of the air forces available, which will be examined later) a plan suggests itself of which the following is a very brief outline.

Intention

To prevent enemy reinforcements and supply traffic from entering the area Bapaume–Le Catelet–Guise–La Fère–Noyon, and to break up the movement towards the battle front of enemy reserves already within that area.

Method

(a) *Reconnaissance.* The British fighter-reconnaissance squadrons on the front of the attack and the corresponding squadrons

---

[1] Compare the supply situation in the Second British Army area in May 1918. See p. 113.

of the French First Army to maintain continuous patrols over the area (incl.) Douai–Denain–Solesmes–Guise–Laon–Soissons, to report all road and rail movement within that area towards the battle front. To the north, the fighter-reconnaissance squadrons of the British First, Fifth, and Second Armies, and southwards those of General Humbert's and Mangin's armies to watch the rail and road communications leading towards the battle area for any abnormal movement. This reconnaissance to be continued at night by the light night-bombers. ·

(b) *Bombing.* The primary task of the bombers—including those attached to the French First Army—must be to cut and keep cut the railways at Cambrai, Le Cateau, Le Nouvion, Vervins, and Laon; the first attacks being made on the evening of the 8th, continued at night and supplemented by the light night-bombers derailing trains seen on the move in any direction on either side of the points mentioned above.

(c) During the first ten hours of the battle day-bombers will attack objectives as follows:

  (i) At Zero hour (from a low altitude under cover of fighter escort) the head-quarters of the German Second Army and of the various corps in that army.

 (ii) The billeting areas of the fit divisions of the Second Army in reserve behind the Somme (the 54th and 243rd).

  Subsequently these divisions to be attacked *en route* to the battle-field by assault aircraft. (It can perfectly safely be assumed that these two divisions would get on the move—as, in fact, they did—as soon as news of the attack got back.)

(iii) Points on road communications mentioned in para. (e) below.

(d) *Low-flying attack.* The maximum number of squadrons that can be spared from duties in connexion with the maintenance of air superiority to be directed mainly against formed bodies of enemy troops and transport in the enemy back areas. Particular attention to be paid to the bottle-necks such as those mentioned on p. 178 and to the Somme bridges.

As a subsidiary operation with the object of assisting the infantry and tanks on to the blue line, the assault squadrons of

the 22nd Wing will be placed under command of Corps as
follows:

IIIrd Corps	Two squadrons.
Australian Corps	Three squadrons.
Canadian Corps	Three squadrons.
Cavalry Corps	One squadron.

These squadrons to revert to command of A.O.C. at G.H.Q.
on the evening of 8th August. Similar arrangements to apply
to the assault aircraft supporting the First French Army.

(e) The unfit divisions (the 107th and 21st) which were known
to have only been recently taken out of the line, were not so
certain immediately to be thrown into the fight. Similarly
there must be uncertainty as to which of the reserve divisions
of the armies on the flanks would be used to reinforce the Second
Army. It would therefore be no good attempting to attack all
these divisions blindly, and the first task of the reconnaissance
squadrons must be to examine their billeting areas and report
on any movement. As soon as any of them are reported as moving they should be attacked by assault aircraft, and by bombers
at suitable defiles on the roads while the columns are actually
passing through them.

(f) Finally, to enable this plan to be carried out, and generally
to ensure for our offensive both in the air and on the ground a
secure base against enemy air action, a high degree of air
superiority must be maintained against the inevitable rapid
concentration of German aircraft over the threatened area.
This requirement can best be met by the vigorous execution
of the air offensive on the lines outlined above, combined with
strong and intelligently directed offensive patrols by the fighters,
to seek out and destroy enemy aircraft in the zone of operations
of the bombers and over the enemy aerodromes.

X

THE R.A.F. IN THE BATTLE (*contd.*)

TACTICAL CONCENTRATION

'The Selection of a correct object demands knowledge and judgement to ensure that the resources which can be made available are sufficient for its attainment.'[1]

THIS factor, which must, of course, be considered before it is possible to arrive at any plan, has not been dealt with earlier in this review because it is one which will repay examination in some detail. It may be admitted at once that the air forces which did in fact take part in the battle of Amiens would not have been sufficient for the attainment of the object which it has been suggested might have been set before the R.A.F. in that battle—in particular, a large number of additional fighters would have been required for low-flying attack. But the wording of the extract quoted above from the 1929 edition of the Field Service Regulations should be particularly noted in this connexion—'*the resources which can be made available*'. This brings us back at once to the key principle of all sound strategy, the principle of concentration, which the writer makes no excuse for again quoting in full:

'(i) *The principle of concentration*: The application of this principle consists in the concentration and employment of the maximum force, moral, physical, and material, at the decisive time and place (whether that place be a strategical theatre or a tactical objective).'

This is followed immediately by a second principle:

'(ii) *The principle of economy of force*. To economize strength, while compelling dissipation of that of the enemy, must be the constant aim of every commander. The application of this principle implies the use of the smallest forces for purposes of security, of diverting the enemy's attention or of containing superior enemy strength, consistent with the attainment of the object in view. This principle is a necessary complement to the principle of concentration.'

[1] F.S.R. ii, sect. 7. 2.

It will, no doubt, be generally agreed that on August 8th the decisive place in the west was the British Fourth Army front. And the application of the principle of concentration, as illustrated by the British air concentration for the battle of Amiens, should be considered under four headings, closely interrelated:

(a) The concentration of the maximum possible number of aircraft on the decisive task, that is to say against objectives directly connected with the attack by the Fourth Army;

(b) The development of the maximum effort by the squadrons so concentrated;

(c) The concentration of that effort upon the minimum number of objectives calculated to give effect to the object selected; and

(d) Centralization of control, with a view to securing due co-ordination and the most economical application of the force available.

In considering first the *concentration of squadrons upon the decisive task*, it is again necessary to remind the reader that in addition to the British squadrons a large number of French air units took part in the battle, comprising not only the units of the French First Army but also those of the General Reserve which were allotted to support them. So that although it is only possible to make a detailed examination of the British figures—which will serve to bring out the point it is desired to emphasize—yet when comparing them with the German forces opposed to them it must not be forgotten that the Allies together disposed of a very much larger force than that of which the details are examined in the following paragraphs.

It will be seen from Appendix C that the British air force in France on August 8th comprised 74 squadrons[1] with a total of 1,390 aircraft serviceable on charge—not counting the 20 army co-operation squadrons with which we are not here concerned. Included in the air concentration for the battle of Amiens are the squadrons of the 5th (Fourth Army) Brigade, most of the 9th (G.H.Q. Reserve) Brigade, and all the 3rd (Third Army) Brigade, the inclusion of the latter being rather more favourable than is strictly justified, because by no means the whole effort

---

[1] Two of these were American squadrons, equipped with British aircraft.

of the 3rd Brigade was occupied by the battle. So the disposition of the British air force in France on the crucial date can be summarized by classes as follows:

	Day bombers		Night bombers		Fighter Recce.		Fighters		Total	
	Sqns.	Air-craft	Sqns.	Air-craft	Sqns.	Air-craft	Sqns.	Air-craft	Sqns.	Air-craft
Concentrated for the Amiens battle	8	147	5	74	3	57	18	376	34	654
Engaged in other sectors . .	10	167	8	121	3	53	19	395	40	736
Total . .	18	314	13	195	6	110	37	771	74	1,390

In effect, this means that only 47 per cent. of the British striking-force squadrons in France were concentrated on the decisive task on August 8th. The German post-War training regulations lay down that 'formations not taking part in the decisive fight must at this time dispense with aircraft';[1] and it is interesting to compare the British concentration for this battle with that of the Germans for their great offensive against our Fifth Army in the previous March—bearing in mind that they were holding more than twice the length of line in France that we were. The total German strength in bombers, fighters, and battle aircraft (i.e. low-flying attack aircraft) *from the sea to Switzerland* on March 21st was about 842 aircraft;[2] of these 482, or about 57 per cent., were concentrated on the front of the three attacking armies. The total concentration on these three army fronts, including reconnaissance and artillery squadrons, was about 730, a concentration at the decisive point approximately equal to the number that was concentrated against them when our turn came to attack in August,[3] although we—the British alone—were then very much stronger in numbers than the Germans on the whole western front, and the Allies between them must have outnumbered the enemy by nearly 3 to 1.

[1] For instance, they withdrew all their aircraft except one squadron from the Italian front for the attack on March 21st.

[2] See Appendix D. The figures given on this page are those of the approximate *actual strength*, which the Germans say was—on the average—only two-thirds of their nominal, or *establishment*, strength; they say that actually many of their units were well below two-thirds of establishment, owing to the difficulty of replacing wastage.

[3] 654 bombers and fighters plus 100 army co-operation aircraft of the 5th Brigade.

Of course in those days we had not been brought up to think of the air arm as a definite strategical striking force. The organization and system of command had grown up from small beginnings on the basis of decentralization to Armies, and it was not easy, especially about this time, to get Army Commanders to disgorge any form of reserves. But another interesting contrast is that between the air concentration and that of the Royal Tank Corps for August 8th. It will be remembered[1] that ten battalions of heavy tanks and two of whippets, making a total of 456 armoured fighting vehicles, were placed under the orders of the Fourth Army for the attack. There were three other battalions in France at the time, but they were still equipped with old tanks of obsolete type and inferior performance, and actually every up-to-date tank in the B.E.F. was concentrated for the Amiens battle.

The disposition and employment of the 736 bombers and fighters which did *not* take part in the battle deserves a very brief examination.[2] Firstly, there were the squadrons under the orders of the three Armies to the north, from Arras to the sea. These included a fighter-reconnaissance squadron with each Army, doing the very important work of medium reconnaissance and photography, which was essential and of which the results might have an important bearing on the operations in the active sector. There were also five bombing squadrons among these three Armies, in addition to two squadrons of the 9th Brigade not engaged in the Amiens battle; and between the 8th and 11th raids were carried out against objectives such as Bruges docks, Bailleul, Menin, Merville, and Estaires, probably mainly in connexion with the enemy's very obvious preparations for an offensive in the Lys sector. And finally, there were the sixteen fighter squadrons engaged in the normal duties of offensive patrol and protection of the artillery aircraft.

On the Belgian coast, in the area about Dunkirk were the five bomber and three fighter squadrons of the 5th Group. This formation, composed mainly of ex-R.N. Air Service units under the orders of the Vice-Admiral Dover Patrol, had been employed in co-operation with the navy against the enemy submarine bases at Bruges and on the Belgian coast, and had taken part in the operations culminating in the attack on

[1] *Vide* p. 152.   [2] *Vide* Appendix C.

Zeebrugge in May. Since that date, however, although the submarine docks and workshops at Bruges, Ostend, and Zeebrugge were still periodically bombed, the bulk of the weight of the 5th Group's attacks had been directed against the bases of the German air force, which had become increasingly active in this sector. Actually, on the 13th of August a massed attack was carried out by six squadrons against the aerodrome at Varssenaere, after considerable rehearsal and preparatory work in the previous week—during which period, it should be noted, the Fourth Army was attacking at Amiens.

The question of the Independent Force has already been examined in some detail.[1] In August this force comprised three day-bomber, one light night-bomber, and two heavy night-bomber squadrons, and was operating at full capacity against the centres of German war industry and communication in the Saar basin, Lorraine, and the Rhineland.

Now these activities, all of them excellent in themselves, appear difficult to justify under the circumstances *at the time.* The bombing on the other Army fronts no doubt was a source of some inconvenience to the enemy, but the operations of only seven squadrons over such a large area can have been little more, nor were the objectives attacked of any vital importance at the time. Again, the withdrawal of some of the fighter squadrons from these Army fronts might have resulted in a temporary set-back to our air superiority and some interference with our artillery patrols in the sectors concerned; but this state of affairs, though irksome, would not have been of vital importance anywhere but on the decisive front; and in point of fact it was a common experience that when active operations were in progress in any one sector, enemy air activity was reduced to a minimum on other parts of the front.

As regards the 5th Group, although a year earlier the campaign against the enemy submarines had been of supreme importance, by July 1918 the submarine menace was virtually a thing of the past, and in practice by that time the 5th Group was mainly engaged against enemy air forces. But though the enemy bombing activity in the north was a serious nuisance, it was no more; and it is difficult to see any justification for the fact that, at a time when every available bomber and fighter

[1] See Chapter V.

was needed for the vitally important operations in front of Amiens, the eight valuable squadrons of the 5th Group were engaging (under the orders of Vice-Admiral Dover Patrol of all people) in a sort of private war against German air bases in Belgium. Similarly the operations of the Independent Force —apart from some value as a diversion—can have had little or no effect on the situation at the decisive point at the time. Some activity on these extreme flanks of the Allied line was certainly justified and necessary in an endeavour to compel the enemy to dissipate his strength in defence. But an equal result could have been obtained by a temporary intensification of effort on the part of fewer squadrons; and in any case it is more than doubtful whether the presence of night bombers either at Dunkirk or Nancy had any effect in containing a superior or even equivalent enemy force opposed to them.

Finally, it is of interest—rather academic interest, perhaps— to reflect that there was in existence another very strong potential reserve in the 400 aircraft of the home defence force.[1] The two Air Defence Areas at home in August 1918 included three squadrons equipped with first-class two-seater fighters (72 Bristol Fighters), two squadrons with light night bombers (48 F.E.2b) and seven squadrons with single-seater fighters (168 Camels) of the same type as those with which many squadrons in France were armed. Many of the fighting personnel were experienced pilots who had already seen service in France. No air attack in England had taken place for nearly twelve weeks, and the units were therefore all fresh and fully up to strength. In reality there were at the time insuperable difficulties in the way of using any of these units in the field, notably the political objections which at that time would undoubtedly have ruled out any idea of their removal from England. The administrative difficulty would have been hardly less serious, since many of the units were split up into detached flights, and few were organized on a fully mobile basis. As a matter of fact a few weeks later in October the question of reinforcing the front in France by units from home establishment was seriously

---

[1] An enormous number of aircraft, comprising about thirty squadrons, were also engaged in coastal reconnaissance and anti-submarine patrols round the English coast. But few of these were of a type suitable for employment in other forms of air operations, though of course they locked up a large number of personnel.

considered.[1] But in August it was not a practical possibility, from which we may perhaps deduce two lessons for the future: first, that there are occasions when the course of action which is strategically the soundest may be absolutely ruled out by political considerations or administrative limitations; and secondly, that air forces should not be regarded—save in quite exceptional circumstances—as permanent fixed garrisons, tied down to the defence of any particular area; but that full use should be made of their mobility to employ them wherever they may be of the greatest value at the time.

Table 2 to Appendix C contains the details of the air force concentration which is suggested as being necessary and adequate to carry through successfully the plan outlined on pp. 172–81. It can be summarized as follows:

	Day bombers		Night bombers		Fighter-Recce.		Fighters		Total	
	Sqns.	Air-craft	Sqns.	Air-craft	Sqns.	Air-craft	Sqns.	Air-craft	Sqns.	Air-craft
Engaged in Amiens battle . .	15	264	13	195[2]	3	57	30	633	61	1,149
Engaged in other sectors . .	3	50	..	..	3	53	7	138	13	241
Total . .	18	314	13	195	6	110	37	771	74	1,390

The advantages of such an overwhelming display of force at the decisive point are obvious, nor can there be any question as to its necessity if the rather ambitious object on which the plan is based was to be attained with a reasonable degree of certainty.[3] So a more detailed survey of the proposal may usefully be confined to an examination of the disadvantages and difficulties that seem at first sight to attend it.

First, under the heading of *mobility*, is the proposal adminis-

[1] Not only by squadrons from home defence but also by others specially raised from training establishments. The writer is not certain to what extent the latter idea had official sanction; but he has a vivid recollection of a somewhat melodramatic conclave presided over by Admiral Mark Kerr—making his first appearance in the new R.A.F. uniform—at the H.Q. of the South-Western Area, when all the officers commanding Flying Training Schools in that Area were told to go away and draw up plans for the immediate formation of service squadrons from their flying instructors and service type training aircraft. The C.F.S., for instance, was to raise one squadron of Camels and one of S.E.5a.

[2] Including five squadrons (57), heavy night bombers of the Handley-Page type.

[3] Especially as in 1918 it was necessary to make up in volume for what we lacked in technical efficiency and precision of bombing compared with present-day standards.

tratively practicable? Could this number of squadrons have
been adequately accommodated, administered, and supplied
in one area? And, moreover, could this concentration have been
effected without forfeiting *surprise*, which was the basis of the
plan for the battle of Amiens?

There is no doubt that the proposal would have been
perfectly practicable from the administrative point of view.[1]
The problem presented by an air force even of this size pales
before the complexity of the vast administrative and supply
organization behind the ground forces of the Fourth Army on the
same occasion, with their 14 infantry and 3 cavalry divisions,
their 2,000 guns and 460 tanks. The country in the neighbourhood of Amiens contains innumerable sites for aerodromes; and,
in point of fact, a few months earlier 45 new aerodromes were
hurriedly prepared in that country within the fortnight from
March 23rd to April 5th, for the accommodation of the squadrons that had to clear out of their aerodromes on the Fifth Army
front before the advancing Germans. But concentration of air
strength does not involve the assembly of vast masses of aircraft
on one group of aerodromes in a circumscribed area—in fact
such a course is obviously undesirable, both from the point of
view of surprise and as offering a very vulnerable objective to
enemy air action. It means merely that the whole effort of all
the squadrons involved is directed under unified control on a
single co-ordinated plan, and its most essential requisite is a
first-class signal organization. Indeed, an air concentration
may be a most valuable means of deceiving the enemy as to our
intentions. By the beginning of August it must have been apparent to the enemy that his preparations for an offensive in
the Lys sector would be known to us; and the appearance of
additional squadrons in the Béthune area, for instance, would
probably have been interpreted merely as a defensive concentration to meet that attack. But the Béthune area is within very
easy range for day bombers of many of the objectives which
they would be required to attack in connexion with the Amiens
battle—such as Cambrai. The great majority of the squadrons

[1] We had already shown what we could do in this line when we were really
put to it. Forty-four out of the fifty-eight squadrons available were concentrated
on the fronts of the Fifth and Third Armies during the German offensive in March
1918.

of the three British Armies on the left, with the exception of those
in the extreme north, could have played their part in the battle
from their existing bases in those Army areas; while the heavy
night bombers with their longer range could have concentrated
at localities well to the rear—such as Abbeville or Étaples—
where their presence, even if located, could have given little or
no indication as to the particular sector of the front where they
were to be employed.

Nevertheless some preparatory moves of squadrons would
have been necessary, such as those from the Nancy area and
from the extreme north—say between Cassel and the sea; and
of these the fighter squadrons, which had comparatively short
endurance, would have needed bases in the vicinity of the battle
area—say in the area St. Pol–Doullens—if they were to develop
their maximum effort.   But all that was necessary in the way of
preparatory work was the clearance of the actual sites and the
establishment of fuel, ammunition, and bomb dumps, which are
easily concealed, and of a nucleus ground organization.   It has
already been suggested that even at that time we were unneces-
sarily dependent on hangars.[1] And if the organization and training of an air force is based on a true appreciation of the importance
of strategical mobility, the only signs of a concentration of air
units should be the presence of aircraft themselves on the
landing-grounds—since the transport can usually be concealed.
On this occasion this evidence need not have been apparent
before Z day.   Squadrons from the north could have been on
their normal aerodromes on the morning of August 8th,
returning to their new battle aerodromes after their first attacks.
And those from the Nancy area could have been moved on the
afternoon of the 7th to staging landing-grounds somewhere like
Épernay—for which a professed anxiety about the situation in
the Lys sector would have provided an excuse if one were
needed.   All this, no doubt, would have involved considerable
organization and very careful arrangements for ensuring
secrecy; but again, seeing that the strength of the Fourth Army

---

[1] See p. 56 above.   The Germans were fully alive to the importance of mobility
for the ground organization of their air units.   For instance, von Hoeppner says
of the bombing squadron formed at Metz in August 1915 that 'with the object
of facilitating its transport from one front to another it was accommodated normally
in military trains'.   See *Germany and the War in the Air*, French translation (Payot,
Paris), p. 75.

on the ground was roughly doubled in the first week of August without betraying the secret to the enemy, there can have been no insuperable difficulty about concealing with equal success a corresponding concentration of air units. Indeed, there seems no reason why we should not have gone farther and aimed at deliberate deception of the enemy on the lines of Allenby's dummy camps and horse-lines in the Jordan valley later in the year, by the establishment of dummy aerodromes in the north and the ostentatious concentration upon them of obsolete or unserviceable aircraft.

There is, however, one point of some importance in connexion with *surprise* which is worth a moment's consideration because it may contain a lesson for the future. Many of the reinforcing units would have been unfamiliar with the battle area. No squadron can be expected to develop its full efficiency if it is required to work over country which few or none of the personnel have ever seen before; and for assault aircraft, especially in close support of troops on the ground, knowledge of the ground is an absolute essential. Obviously the sudden great increase of air activity that would have resulted if all the reinforcing squadrons were allowed to fly over the area to get to know the country must have aroused the enemy's suspicions, and though much could be done by the study of photographs, that in itself would not be enough. So some arrangements would be necessary on the lines of those made to conceal the registration of all the reinforcing artillery, in the form of programmes 'carefully worked out giving the times at which guns should fire and the number of rounds to be fired so that even though the amount of artillery in the line had been doubled, the enemy should not appreciate it'.[1] Probably the simplest arrangement would be to attach squadron commanders, patrol leaders, and as many other officers as possible from reinforcing units, to units of the 5th Brigade during the period before the battle, their place being temporarily taken in their own units by officers from the 5th Brigade who already knew the battle area.[2]

The concentration suggested may be criticized on the grounds of *security* as weakening to a dangerous degree the air defences

[1] *The Story of the Fourth Army*, p. 20.
[2] The Germans did something on these lines before their great attack on March 21st, 1918. See *Military Operations, France and Belgium, 1918*, p. 155.

of other sectors, or on those of *economy of force* as leaving inadequate air forces to compel the enemy to dissipate his strength in defence.   To such criticisms the answer is, firstly, that air forces are not irretrievably committed to any task or any sector, but by virtue of their inherent mobility could have been switched back at a moment's notice to any point where a threat had arisen so urgent as to require their presence, and to outweigh the importance of their influence in the battle area. Secondly, although the squadrons remaining in other sectors as detachments would have had to redouble their activities to compensate for their lack of numbers, it is nevertheless doubtful whether in these circumstances they could have selected any objectives so vital to the enemy as to contain superior or even equal forces for their defence.   But on the basis proposed the Allies would have disposed at the decisive point a force so immensely superior in strength to that of the Germans, that the utmost concentration which the latter could effect must still have been heavily outnumbered.   Moreover, an air offensive on this scale must have had the effect of forcing a defensive attitude upon the enemy, and it cannot seriously be maintained that the Germans would have been capable, with the forces at their disposal, of any serious counter-offensive in the air elsewhere along the front.   Some small risk there might have been, but the power to achieve great results in war entails the reasoned acceptance of calculated risks, and this situation—as nearly always in air warfare—would surely have proved again the supreme value of offensive action.   Moltke, in 1914, weakened his decisive right flank in order to provide against minor risks elsewhere—and thereby probably lost the war; so in the air in 1918 a cautious policy of leaving small detachments to provide against minor contingencies in less important sectors must have fatally weakened our punch at the decisive point.

Turning now to the second factor in concentration, that is, the *development of the maximum effort by the squadrons engaged*.   British fighting policy both on the ground and in the air throughout the War was to exert a constant pressure on the enemy.   Unlike our French allies, whose tendency was to fight like tigers when there was a battle in progress, but, when there was not, to live and let live, the British Command discouraged 'peace sectors' on the ground. And unlike the Germans, who deliberately restricted flying and husbanded their resources during quiet periods, our air force never really slacked off, and it was almost unheard of for a squadron to be taken out of the line to rest or reserve. It may perhaps be for this reason that the squadrons do not seem to have operated during the first crucial four days of the Amiens battle at anything like their maximum capacity. The 9th Brigade, it is true, had been engaged in the fighting on the Marne, in which it had done very hard work and suffered considerable casualties; but nearly three weeks intervened between the Marne battle and the 8th of August, which should have been enough to rest the 9th Brigade, and all the squadrons concerned were well up to strength in personnel and aircraft. Nevertheless it will be seen from Appendix B, Table 2, that, taking as a 100 per cent. standard the actual performance of one squadron, No. 205, which for the four days, 8th to 11th, did an average of two tasks per day for each aircraft on charge, the expenditure of effort by the other bombing squadrons fell extremely short of that standard, and in fact the average was only just under 50 per cent. of the maximum. This performance is in striking contrast with the activities of many of the same squadrons during the German offensive in the previous March, when, for instance, No. 101 squadron dropped no less than 502 25-lb. bombs on Cambrai and Ham in a single night, which must have represented at least three raids per aircraft for a considerable proportion of the squadron.[1] The effort that we put forth to avert disaster to our own arms should at least be equalled by that exerted to inflict disaster upon the enemy. During the less critical stages of a campaign flying should be to some extent deliberately restricted—for instance during those 'periods of comparative inactivity' on the land front when the bombers may be undertaking operations against enemy production. Whole squadrons should be taken out of action into reserve for respectable periods of rest and refit, and the *tempo* of the operations should be reduced to within reasonable limits. Airmen cannot be expected to fight every day for months on end—any more than infantrymen. When a great battle is known to be imminent the personnel of all units should be

[1] See also *The War in the Air*, vol. iv, p. 385, for the performance of this same squadron during the German Lys offensive in April.

rested as much as possible consistent with maintaining the necessary air situation. But when once that battle is joined, then every nerve and sinew of our personnel and every resource of our material must be strained to the utmost—in fact, we must concentrate 'the maximum force, moral, physical, and material, at the decisive time and place'.

The next consideration is that of *concentration of all the available effort upon the minimum number of objectives calculated to achieve the object.* Marshal Foch's instructions to the Allied air forces, already referred to,[1] laid the greatest emphasis on what he described as 'the *essential* condition of success . . . the concentration of *every resource* of the British and French bombing formations on such few of the *most important* of the enemy's railway junctions as it may be possible to put out of action with certainty, and keep out of action', and during the German offensive he had allotted to the whole of the British and French air forces and the Italian Eastern Air Detachment a total of only *eight* railway objectives.

In the twenty-four hours beginning at Zero on the 8th of August, when only eleven British bombing squadrons were in action, twenty-two[2] objectives were attacked on the Amiens front alone, of which only nine were attacked more than once. On the plan suggested twenty-eight British bomber squadrons would have been in action on the first day. It is a little difficult to say how many objectives would have been attacked during the first ten hours, but counting the billeting areas of the 54th and 243rd divisions each as one objective, and including the various head-quarters and the five road bottle-necks mentioned on p. 178, the maximum number should have been about a dozen distributed among all the Allied bomber squadrons. By the evening of the 8th the bombing would have shifted on to the railways, and from then onwards the bulk of the effort of twenty-eight British and all the available French squadrons—which together could have amounted to little less than fifty, comprising probably about 700 bombing aircraft—would have been directed mainly against five points on the railways. It does not seem impossible that even in the conditions of 1918, the con-

[1] See p. 127.
[2] This is *not* including a certain number of objectives which were bombed by single aircraft only.

tinuity and intensity of attack represented by a force of this size might have completely stopped rail traffic through those five points; and the six divisions from the north that came through Cambrai must at least have suffered heavy casualties, and probably a delay of two vital days.

Finally, we come to the question of *centralized control of air operations*. Here we meet at once with the difficulty, with which we may again be faced, of any effective unified control of inter-allied forces. In point of fact unified command—real command—by an Allied general over a mixed force never has existed and pretty certainly never will: Marshal Foch has described himself as 'the conductor of an orchestra', and that is probably the limit to which unified command can ever go. Some arrangement on the lines suggested in a previous chapter—the appointment under the Generalissimo of an Air Officer Commanding an Inter-Allied Air Reserve and charged with the co-ordination of the Allied air effort in France—would have been of great value on this occasion. Actually, of course, Débeney's First Army was placed by Foch under Haig's orders for the Amiens battle. The British General Staff issued to Débeney a copy of the R.A.F. head-quarters instructions for the battle (themselves very brief and general) with a request that he should co-ordinate the action of the air forces under his command with the action being taken by the R.A.F. But this and the definition of recon-naissance boundaries was the extent of the co-ordinating action taken, and there was no more detailed allotment of tasks such as objectives to be bombed or fighter patrol areas. Probably in the circumstances at the time nothing more was practicable.

Within our own Service the control of the air striking force in France was decentralized among eight different commanders, from the Vice-Admiral Dover Patrol in the north, through the five different Army Commanders, to the A.O.C. Independent Force at Nancy in the south, with a reserve of sixteen squadrons in the 9th Brigade under the hand of the A.O.C. at G.H.Q. At the battle of Amiens the commanders concerned were the 5th Brigade (two bombers and eight fighters) under the orders of the Fourth Army Commander, the 9th Brigade, who received their orders direct from the A.O.C. at G.H.Q., and the 3rd Brigade (two bombers and four fighters), who were under the orders of the Third Army but who also (rather curiously) received orders from the A.O.C. at G.H.Q. In addition, one squadron from each of two other brigades also took part in the battle, presumably on the orders of the A.O.C. at G.H.Q. But disregarding these last, and including the French air force as under one commander (in point of fact it was probably under at least two), it still remains that four different commanders were concerned in the direction of the air effort in this one battle—subject to a certain amount of co-ordination by British G.H.Q. The separate French commander was probably inevitable, but there appears to be no strong reason against centralizing the control of the British air forces under a single officer. On the ground one British General was fighting the battle—General Rawlinson—and he and his staff should surely have had to deal with one single air commander and one only. On this occasion, if the plan suggested had been adopted, the bulk of the R.A.F. in the field would have been engaged; and the control of the air operations must then have been exercised by the Air Officer Commanding from G.H.Q., with an advanced battle head-quarters near Amiens, in close contact with the head-quarters of the Army Commander fighting the battle, and in direct telephone communication with the various subordinate air commanders, the meteorological centres along the front, the Intelligence branch at G.H.Q., and the head-quarters of the Aviation of the First French Army. Only by some such means could the air situation on the entire front have been surveyed as a whole, and the combined efforts of all the Allied air striking forces have been properly co-ordinated and directed to the attainment of the one common object.

There can be no value in a more detailed elaboration of the plan suggested on p. 179. The exact employment of the various squadrons, the proportion of the total strength allotted to the various tasks, and the actual objectives attacked must have depended upon circumstances as they arose, upon the information obtained by reconnaissance, and upon the degree of success attained by the early operations. The general lines of employment of the bombers have already been indicated with sufficient clearness. Of the thirty squadrons of fighters as high a proportion as the air situation would permit should have been engaged in low-flying attack; probably on the opening day more than half that number, including all the squadrons of the 5th and 9th Brigades, would have been employed, some on the battle-field in the direct support of the infantry on to the blue line, and others beyond the Somme in attack on the first reinforcing divisions approaching the area by road. The remaining fighters, amounting to about 300, would have held the ring by offensive patrols over the enemy aerodromes, over the zones where the assault aircraft were engaged, and over the bombers' objectives in direct co-operation with the massed raids that would have been in progress throughout the day.

One word of warning is necessary in connexion with the low-flying attack in this battle. It is unsafe to accept the method of employment of the fighters in August 1918 as a basis on which to form our ideas of their future employment—especially in the earlier stages of a war. At this time both the Allied High Commands were deeply impressed with the great services that had been performed by the low-flying fighters during the defensive battles of March and April. It will be remembered that Foch had laid down in his instructions of April 1st[1] that low-flying attack was the 'first duty of fighter aeroplanes'. Unquestionably their employment in this role was fully justified in the circumstances at the time. But at the time of the battle of Amiens we had in France alone more than three times the number of fighters than we have in our whole air force in 1935, including the Fleet Air Arm; and we had also an enormous numerical superiority over our enemies. This point has been fully dealt with elsewhere, and it is unnecessary to labour it more than to emphasize once more that, although air superiority is only a means to an end, it is an essential factor in the success of any air plan; and that while we have only a comparatively small number of fighter squadrons available it will seldom be possible —unless the enemy air opposition is negligible—to divert those squadrons from their more normal task of seeking out and destroying enemy aircraft, keeping the ring clear for our bombers and army co-operation aircraft to go about their normal occasions with the minimum of interference.

Finally, it may be of interest to describe briefly what the enemy did to reinforce his Second Army front between the 8th and the 11th. Unfortunately the writer is not aware of what routes the

[1] See p. 105.

German reinforcing divisions actually did use, but the dates on which the various divisions were identified by prisoners taken by the Fourth Army are known,[1] and from this and the known positions of their billeting areas before the battle it is possible to deduce with reasonable accuracy the way they came.

Taking them in the same order as on p. 176, the 43rd and 109th divisions shown as being at Flaucourt and Berny were in fact caught by the attack in the act of relieving the 108th and 117th divisions in the line, and were roughly handled by the low-flying fighters of the 22nd Wing. The 54th from Roisel and the 243rd from Ribecourt were identified on the 8th and 9th respectively, so that they evidently did get on the move to the front without delay and by the shortest route; the 54th probably crossed the Somme at Péronne and Brie bridges, and possibly also at St. Christ, moving thence by Estrées and Fouceaucourt; the 243rd moving by Gouzeaucourt and Fins, thereafter probably followed the same routes. Of the two tired divisions east of Le Catelet, the 107th was not engaged during August, and the 21st Division not until the 13th; so reconnaissance should have revealed that there was no big movement from the Le Catelet area early in the battle.

Of the fit divisions in reserve in the areas on the flanks of the Second Army four were identified on the Fourth Army front by the 11th. The 82nd Reserve from about Guiscard moving probably by the road Nesle–Chaulnes was identified on the 9th. On the next day the 5th Bavarian and the 221st divisions from the Seventeenth Army were identified; of these the 5th Bavarian from the Marquion area must have got on the move early on the 8th and probably came by road—about thirty-eight miles—via Péronne; while the 221st north of Denain was over fifty miles away and is more likely to have come by rail via Cambrai. The 26th (Wurt) Reserve, also of the Seventeenth Army, was south of Bapaume when the battle began, but actually was not in action against the Fourth Army till the 11th. In addition there was a less expected reinforcement between the 8th and the 11th, in the form of two divisions, the 1st Reserve and the 204th (Wurt), which were reported as being still in the line in the Eighteenth Army area to the south.

Finally, six more divisions arrived and were identified by the

11th coming from still more distant areas, and of these—in accordance with expectation—five came from the north, almost certainly by rail via Cambrai. The 79th Reserve from about Roubaix, the 121st from Courtrai, and the Alpine Corps from south of Ghent all belonged to the Fourth German Army in the northern end of the line and were identified on the 10th, 10th, and 11th respectively; while the 119th from about Tournai identified on the 9th, and the 38th from Lille on the 11th, were both drawn from the Sixth Army behind the Lys salient. The one division from the south, the 10th division from the Seventh Army, came from about Laon and evidently moved by rail via La Fère to be identified in action on the 9th of August.

Thus in all reasonable probability six divisions—comprising between two and three hundred trains—passed through Cambrai, most of them within the forty-eight hours from the evening of the 8th to the evening of the 10th of August—an average of about one train every fifteen minutes.

PART IV
CONCLUSIONS

XI
THE THIRD REVOLUTION

FROM time to time a new invention astonishes the world, and is hailed by the prophets as the forerunner of a revolution in the military art. The cross-bow, the rifled barrel, the quick-firing gun, the submarine, the railway, and the motor-lorry—all these and others in their day have forcibly imposed important modifications in technique, and wrought great changes on the face of war. But all of them have had their counterpart in earlier ages, and none can really be said to have changed the nature of war. Even the vast field fortifications of the Great War were nothing new: the same fields were scarred by the trenches dug by the men of Marlborough and Villeroy in the eighteenth century as by those of French and Falkenhayn in the twentieth. And that the guile and manœuvre that served Marlborough to force the lines of Brabant had to be replaced for Haig by the heavy artillery and tanks that pierced the Hindenburg line, was in large part due to one of the three factors which alone, perhaps, deserve the title of revolutionary. Of these the first in point of time was the invention of gunpowder. The second, the machine-gun, seems likely to be cancelled out by yet another, as armoured fighting vehicles prove their worth—unless it can hold its own by giving birth to a really effective light, mobile, anti-tank weapon; for the moment the machine-gun petrifies manœuvre, and holds land tactics in a grip like paralysis. The third revolution is the conquest of the AIR, and this is the most revolutionary of all. For where other weapons have enhanced the capacity of men to kill each other in battle, and increased the depth of the battle-field, the AIR may stop men or their supplies arriving at the battle-field at all. That the advent of three-dimensional warfare has had even more revolutionary implications in a wider sphere is well known; but these lie beyond the scope of this book, which is restricted to an examination of the influence which air power may exercise in a campaign on land where armies are engaged. Our actual experience in this field is limited to that of only one major war between civilized Powers; and it is for this reason that the writer has hitherto endeavoured to refrain from any dogmatic statements or definite claims for the efficacy of air action against armies, but to confine himself to suggestions as to what its effects *may be*, as a basis for further thought and experiment. It seems desirable, however, to accept the risk of being written down at some future date as a false prophet, and to sum up this book by setting down a few conclusions—deductions, perhaps, would be a better word: deductions are always open to be falsified in one direction or the opposite, and we cannot *know* about these things with so little experience, we can only make a reasonable estimate.

If there is one subject more than another on which an estimate of the effects of air action seems unlikely to be proved false, it is that of its *influence on the maintenance of modern armies.* History teems with examples of defeat owing (at least in part) to defective arrangements for transportation and supply. Massena's campaign before the lines of Torres Vedras; the Russian campaign against the Japanese in Manchuria in 1905; our own early operations in Mesopotamia; the First Russian Army before Tannenberg; even the great German turning movement through Belgium in 1914—all these are instances where failure was due, in a greater or lesser degree, to maintenance difficulties. The role of the air will be to create those difficulties where they do not already exist, and to intensify them when they do. Colonel Wingfield sums up his article, already quoted in a previous chapter, with the following words:

'It does not seem to me impossible that action from the air could definitely prevent trench warfare occurring again as we knew it in France. The war on the Western Front was fought in a country at least as well supplied with railways as any probable theatre of war in the future; and if in that country a situation could arise where the mere maintenance of an army required three-quarters of the maximum capacity of the railways behind it, it is clear that the margin of safety is never likely to be great. It must be remembered that unless the services of maintenance can be kept in working order, the concentration of large armies, highly equipped with modern weapons, is out of the question.'

That article was written in 1925 as a result of the experience of 1918: to-day we can make bold to be more definite. These conclusions are presumably with reference to 'large armies highly equipped with modern weapons' in the sense of the 'million armies' of 1918 with their colossal forces of artillery —they would be less true of armies equipped with really modern weapons as we understand the term to-day. But it is surely safe to go farther and as our *first conclusion* to assert, at least, that the margin of safety on the line of communication of a national army on the man-power and shell-power basis, in the face of modern air action, would be such that no insurance company would consider it as a reasonable risk. It is of some significance that the Army of 1918, on the experience of which Colonel Wingfield bases his conclusions, was approximately the same size as the British Regular and Territorial Armies of 1935 at war establishment.

If this estimate be true of an army operating in a country so well equipped with railways and roads as northern France, with how much more force must it apply to one situated at the end of a *single line of supply*. In two of the examples quoted above the maintenance difficulties which proved so fatal were due to the existence of only one line of supply—the river for Townsend in Mesopotamia, the Trans-Siberian railway for the Russians in Manchuria. Reference has already been made to the possibility of a British army having again to operate beyond the Khyber in the future, as it has done more than once in the past; and there are many other instances, historical and potential. In 1915, after our submarine and air action had to all intents and purposes closed the Turks' sea line of communication to Gallipoli, the enemy divisions holding the Peninsula had to rely practically entirely for supply and reinforcement upon the one road through Keshan and Bulair, within very easy reach of British air bases on Imbros. It is estimated that between May and December 1915 some hundred thousand pack-animals and two hundred thousand troops passed over that road.[1] Could the Turks have held Gallipoli to-day? Or would it be safe to-day for an army of the size of that which operated in Palestine in the

[1] See *The War in the Air*, vol. ii, p. 65.

last two years of the war to depend on one line of railway back to Kantara, supplemented by long and vulnerable camel convoys? Or upon the one pipe-line which carried the water forward from the Sweet-water Canal, and which the enemy did, in fact, make several efforts to cut? In a wider sphere, that most vital of our imperial communications, the Suez Canal, has already been an objective for our enemies in war; and the advent of air power has given added importance to our occupation of Palestine as a factor in imperial strategy—to give depth to the air defence of that very vulnerable artery. There exists to-day a potential theatre of war which may before long provide an instructive example of this aspect of air warfare, namely the Far East, where a clash between Japan and the Soviet is always a possibility. The Japanese profess grave anxiety about the threat to their very vulnerable and inflammable cities represented by the Russian air bases in the neighbourhood of Vladivostok—a threat which, if not very imminent at the moment (for after all about seven hundred miles of sea separate Vladivostok from Japan), may become more real as the effective range of bombing increases. The Russians on their side, partly on this account and partly because of Japan's traditional tendencies on the Asiatic mainland (and of certain other factors such as oil interests in Saghalien), are equally concerned for the safety of their maritime province, the port of Vladivostok, and the maintenance of their very large garrisons on the Amur frontier and in the Vladivostok area. They are doing all they can to make their garrisons self-supporting by such measures as the settlement of reservists in the maritime province, the development of agriculture, and even of local manufacture of munitions. But the safety of Vladivostok cannot fail to be a source of real anxiety to them, lying as it does at the end of one immensely long line of the Trans-Siberian railway, of which many miles are single track only, and all east of Chita are within easy range of potential Japanese air bases on the soil of Manchukuo. They are now engaged in converting to double track several hundred additional miles of this railway and—it is believed—constructing another line farther north, from Lake Baikal to the Pacific. Furthermore, this factor of air action against a single line of communication may well rule out the possibility of military adventure by either side against the other on the ground of Manchukuo. For instance, a Russian advance towards Harbin would be dependent for maintenance upon about 500 miles of the one railway between Chita and Tsitsihar—through a country with few roads and those quite unfitted for heavy military traffic. And although south-east of Tsitsihar the Japanese would have the advantage of three lines of communication—the South Manchurian railway, the Chinese Eastern railway, and another now building from Sishin on the Sea of Japan—an offensive westwards, for instance with the object of cutting off Vladivostok by seizing the junction at Chita, would be subject to the same restrictions as a Russian invasion in the opposite direction. In fact it seems inevitable that any campaign in northern Manchukuo must soon result in deadlock, to be decided by whichever belligerent can force the other's railheads farther and farther back by cutting the communications behind them, while at the same time protecting his own; and in this respect the Japanese in defence would have an important advantage by having three lines to the invader's one. Here then is a *second conclusion*—that, in a future war against a highly equipped enemy, no army can afford to be dependent on a single line of supply within hostile bombing range. The disability suffered by a belligerent whose essential sources or avenues of supply are concentrated in a relatively small number of vital centres has already been noted in an earlier chapter,[1] and a single line of supply in the field is the supreme example of that disability.

A *third conclusion* to which the above considerations also lead is that in future staffs must think wider and use larger maps. The most effective, if not the most immediate or obvious, way in which an air force can assist an army may sometimes be by operating from bases far distant from the army's theatre of war, against objectives on the enemy's line of communication a long way from the front—in effect, by attacking his communications from a flank. Thus the Japanese air force might very likely be most effective in forwarding a land investment of Vladivostok by cutting the railway at Chita from bases near the western frontier of Manchukuo, hundreds of miles from the front. In the Great War the Turkish railway line of communication serv-

Change of Picture

[1] See p. 18.

ing their armies in Palestine, Iraq, and the Hejaz passed along
the coast of Asia Minor into northern Syria; and under modern
conditions British bombers could probably best and most
economically have helped Allenby in Palestine, and Maude in
Iraq at the same time, by operating from bases in Cyprus, within
easy bombing range of objectives on the railway such as Muslimieh junction or the Taurus defiles.[1]

Next we must consider the influence on the *strategic concentration of armies*; first, in relation to the movement to the area of
concentration at the outset of a campaign.   In the last war in
western Europe the rail movement involved in mobilization
and concentration assumed the most staggering proportions.
In 1914 the Germans used 11,000 trains to concentrate on all
fronts over 3 million men and nearly 900,000 horses; their
concentration in the west was at the rate of 660 trains per day,
of which 550 per day crossed the Rhine bridges westward.   The
French concentrated 1,200,000 men and 400,000 horses in
about 4,000 trains—the total traffic on the Est system in the first
16 days of war amounting to about 20,000 trains, including
returned empties and essential civil traffic.   These almost
astronomical figures suggest enormous possibilities of dislocation.
Elasticity is essential if the railway operating staff are to be able
to meet emergencies, and a system is at its least elastic when congested; so there can be no doubt that the results of air action at
this time might have been very damaging to the effective execution of those tremendous programmes, prepared in detail in peace,
which, even as it was, put a very severe strain on the railways.
On the other hand, we are here considering a frontier which on
either side is more abundantly served with railways than any
other area in the world.   On the German side, for instance,
there were fifteen bridges over the Rhine, and thirteen double
track lines led to the detraining areas west of that river; the
French had ten *lignes de transport*, and on both sides the main
lines were supplemented by a complicated network of subsidiary
and branch lines.   There is some doubt whether even in 1914
the pre-arranged schedules were not more rigid than they need
have been, while by this time fresh construction and improved

[1] As a matter of fact something on these lines was suggested by the Military
Representatives at Versailles in a joint note on 9th Jan. 1918.   See *Military
Operations, France and Belgium, 1918*, p. 59.

technical efficiency have probably increased the capacity of European railway systems. Moreover, this is a direction in which the rise of road transportation and the increasing mechanization of first-line transport and artillery have already had a most valuable influence. Even if we were to see again armies of such colossal numerical strength—which is less than likely—modern transportation, regarded as a whole, is more capable of dealing with them, and the railways would not be operating so near their limit of capacity. Road transport has greatly increased the flexibility of military movement; and although it is at least doubtful whether first-line transport or tracked vehicles—especially the latter—can ever be detrained actually out of bombing range, still they will be able to move under their own power much greater distances than horsed transport and artillery; with the result that detrainment areas will be much deeper, many more stations and sidings will be available, and there will be much less vulnerable accumulations of troops, horses, and transport at detraining points than there were in 1914.

These, however, are the conditions in a highly developed part of the world where the railway systems, already naturally prolific owing to the existence of large industrial areas, have for years been multiplied beyond economic need by lines constructed specially for strategic purposes. In less advanced countries, even in Europe, where railways are fewer, less efficient, and less well supported by good roads, the situation would be very different. For instance, the line from Budapest to Belgrade has a very limited capacity—there are some gradients which reduce running speed to about four miles per hour; as it was it took six weeks to concentrate three corps by this line. And it seems fair to suggest that to-day a rapid concentration of an adequate air force against movement on this line might have saved Serbia from being overrun.

So the _fourth conclusion_ that suggests itself is this: that, with air forces of the strength we are reasonably likely to see in the near future, the initial concentration even of large 'man-power' armies cannot be stopped in country as well served by railways as the Franco-German frontier, but can be delayed. On this point the Field Service Regulations[1] state with unaccustomed

[1] See F.S.R. ii, sect. 18.

force: 'It is essential that the strategical concentration should be completed without serious interruption from the enemy . . . any failure or miscalculation in the preliminary work of transporting or maintaining the troops may be fatal, for loss of time at this juncture may lead to loss of the initiative.' As a matter of fact it is doubtful whether this element of time is really of such capital importance in these days, anyway in European warfare, when great permanent frontier fortifications are in any event certain to impose considerable delays not long after armies have left their concentration areas. And our conclusion also finds support from the fact that it is very unlikely that in this early stage of a campaign air forces will as yet have been able to bring their full effort to bear on the enemy's railways. It is true that by the time great armies are due to complete concentration the air forces will already have been in action, probably for a matter even of weeks; but it cannot be supposed that by this time a definite superiority will have emerged, sufficient to free either side from the necessity of diverting at least a considerable proportion of their striking force to deal with the enemy air forces —with a corresponding reduction in the scale of attack available to interfere with the enemy's concentration on the ground.

On the other hand the *fifth conclusion* can hardly be denied, that in a country where communications are scarce and, of course, above all where the enemy's initial concentration depends upon a single line of railway, it is probable that it can actually be prevented. And if, as is by no means inconceivable, the enemy concerned is a third-rate Power with a weak and inefficient air force, the chances amount almost to a certainty.

If and when the movement to the area of strategic concentration has been successfully completed, it seems certain that that concentration will in future have to take a different form from the huge agglomerations of troops in comparatively small areas that were the practice in the past. In that respect we may be able to take a leaf out of the book of naval methods in the last century —Nelson's idea of concentration was, not the assembly of all his fleet in one small area, but the capacity to reach the decisive point with superior force in time, conferred by a good service of information. This then is the *sixth conclusion*—that, in place of our traditional form of disposition within the area of strategic concentration, we must develop a new technique in which

Nelson's frigates and the mobility inherent in naval forces may find their counterpart in modern methods of reconnaissance, intercommunication, and movement.

The forward movement from the area of concentration is probably the first stage, in a highly developed theatre of war, in which the delays imposed by air action may begin to have results of primary importance. This movement, in the nature of things, is bound to be a race against time and against the opposing army, often for some natural feature—a natural defensive position like the Meuse line or a defile like the gap between the Ardennes and the Dutch frontier of the Limburg Appendix. For instance, von Schlieffen in his plan for the invasion of France, which in an emasculated form was put into effect by the younger Moltke in 1914, postulated that success of the huge enveloping movement by the Right Wing depended on the heads of columns reaching the line Ghent–Mons–Montmédy by the twenty-second day after mobilization—an objective which was in fact attained. This initial onrush of an invading army of the traditional model involves the presence of enormous columns on the roads; and a moment's reflection on the air targets that would have been presented by, for instance, the twenty-six divisions and five cavalry divisions of the German First and Second Armies in Belgium in August 1914, must surely carry the conviction that the Defence has in the air striking force a weapon of the utmost strategical value in the opening moves of a war. As an invading army moves forward from its bases it will become more and more susceptible to air attack, until a stage is reached when its maintenance at the end of long and necessarily limited lines of communication will probably become the most profitable objective. This subject has already been discussed earlier in this chapter, but there are two points about it which are more suited to this context. The first, which is particularly applicable to the defence in the opening stages of an invasion, is that the efficacy of air attack on the communications of the invading army can be enhanced by a carefully co-ordinated programme of strategic *demolitions*; the intensity of air attack can be concentrated by reducing the number of vital objectives against which it has to be directed—in effect, by canalizing by means of demolitions the invader's line of communication and supply, and thus increasing its susceptibility

to interference. We have already noted the results of the
demolitions about Liége in canalizing the communications of
the three German right wing armies in August 1914. And it is
an interesting fact that a year later, at the battle of Loos, the
plan of attack on the communications behind the enemy's front
included—in addition to the bombing programme already described—instructions to the French secret agents in occupied
territory to do their utmost to cut the German communications
at various points north of the line Orchies–Valenciennes–
Namur.[1] The second and similar point concerns the action of
armoured forces. A role of increasing importance for armoured
forces is the semi-independent task of harassing the enemy's
communications close behind his front—though the tendency
discernible in recent years for the tank to usurp the long-range
functions of the aeroplane must be guarded against. Now the
risk of raids by fast-moving armoured and mechanized forces
demands that the various installations on a line of communication like forward depots, refilling points, and reserve billeting
areas should be concentrated in as small an area as possible.
This is obviously essential if an inordinately large proportion of
an army's fighting strength is not to be absorbed in protecting
its own tail, and if the necessarily limited supply of anti-tank
weapons is to meet demands. It is thus apparent that the tank
threat to a line of communication pulls in exactly the opposite
direction to the air threat, which demands dispersion.[2] The
moral is that these opposing tendencies should be deliberately
exploited and the two arms, the air striking force and the
armoured force on the ground, should be used to play into each
other's hands—the tanks by raiding the enemy's back areas
must compel him to concentrate his maintenance and supply
installations, and thus create excellent and vulnerable targets
for the air force, and vice versa.

Therefore the *seventh conclusion* is that the forward movement
from the concentration area of an army on the 'early twentieth
century' model may be seriously delayed, conceivably even to
a fatal extent; but that the most dangerous situation for
an army is when it is well forward at the end of long lines of
communication; and that in that situation the action of
every agency of attack on those communications—air, tanks,

[1] See *The War in the Air*, vol. ii, p. 127. [2] See p. 118 above.

P

and demolitions—must be correlated and co-ordinated on a definite plan.

On a rather lower plane of strategical conception we must consider the influence of air action on what Mr. Winston Churchill has called *'that larger form of tactics or battle-field strategy which manifests itself through the adroit use of a superior rail-way system'*,[1] of which the most striking examples are to be found in the history of the German operations against the Russians in the early months of the Great War.[2] The first of a sequence of tremendous strategical moves by rail was made after the battle of Gumbinnen, when the 1st Corps and 3rd Reserve Division were switched by rail to an area 130 miles to the south-west, for the concentration before Tannenburg. On a still larger scale was the concentration for the advance on Warsaw in September when four corps, a cavalry division, and army troops constituting the new Ninth German Army were transported in 765 trains from East Prussia to Upper Silesia. Following the disengagement and withdrawal from Warsaw, the Ninth Army was switched in 818 trains round the nose of the Polish Salient and reconcentrated in eight days about Thorn for the fresh offensive which terminated in the victory at Lodz in November. And early the following year 508 trains moved five corps within a week—one from Poland, one from France, and three newly raised from Germany—into East Prussia for the campaign which ended in the annihilation of the Russians in the forests of Augustowo. It is small wonder that Falkenhayn claims in his memoirs[3] that the railways 'became in effect the re-duplication of the reserves'. Note, however, that all this was only made possible by the constant foresight, vigilance, and efficiency with which the control of the German railways was exercised by D. R. Ost and his subordinate staffs;[4] and that the whole of the railway system by which alone that series of great strategic manœuvres was achieved (including the head-quarters of D. R. Ost) was within easy bombing range—by modern standards— of Russian air bases in Poland. Even as it was in 1914 these moves imposed a tremendous strain on the railways, which were

[1] See *The World Crisis, Eastern Front.*

[2] For details see the article by Major C. S. Napier, R.E., referred to in a footnote on a previous page.

[3] See *G.H.Q. 1914-16*, by General Erich von Falkenhayn, Chief of the German Great General Staff 1914-16, p. 41.   [4] See p. 137 above.

worked up to, and beyond, their maximum capacity. On several occasions the strain came near to breaking-point, due to circumstances of a kind which in any event are always liable to be present in war—refugee traffic fleeing before the Russian invasion, the evacuation of prisoners after Tannenburg, hasty construction and repair, overloading of telephone lines, and normal operating accidents such as derailments and the collision at Könitz already mentioned. And it is impossible to believe that the system could have continued to function under the additional stress of constant dislocation from the air. The history of the Great War contains other examples of movements by rail of equal strategic significance, though perhaps not of comparable intensity—such as those of the French divisions from various parts of the line to the area north of Paris, which went to constitute Maunoury's Sixth Army at the battle of the Marne. And the *eighth conclusion* is that in future the railway can no longer be regarded as an instrument of major tactics; because rapid movements of troops on a large scale from one part of the front to another will no longer be safe, or even practicable, within effective bombing range of a powerful air force; and will therefore take much longer, and allow ample time for the enemy to obtain warning and adopt fresh dispositions for defence.

Turning now to the share of *the air striking force actually in battle*, it is not proposed to discuss here at length a subject which has already been examined in detail in Part III of this book, from which readers may draw deductions of their own. It is, however, necessary to refer to two general points which appear to be of primary importance. The first is that in air action we may have the key to the solution of a problem which must have been exercising the minds of all soldiers since the War—namely the problem of how to convert the 'break-in' into the 'break-through'. This is not merely a trench-warfare problem; it applies equally to any form of military operations on land, and amounts to the question of how to maintain the momentum of an attack—how to carry it through to a decision before the enemy has had time to move his reserves to the threatened point. Various solutions have been offered, mainly in the direction of endowing the attack with additional hitting power and mobility by the provision of armoured fighting vehicles, and by the motorization or mechanization of the traditional arms. But a more mobile form of attack may be met by a correspondingly mobile reserve in defence. It is perhaps too early to assume that even the tank has as yet broken down the supremacy of the defensive on land; and the initial crust may take longer to break into—the field fortifications of to-morrow may be as much superior to those of the last war as the Hindenburg line was to the primitive trenches of South African days. Any measures taken on the ground to enable the attacker to move faster must surely prove inadequate unless they are supplemented by measures from the air to make the defender's reserves move more slowly—or prevent them moving at all. We cannot afford to assume that our fast-moving, hard-hitting force on the ground will be able to carry the assault through to a decision before the enemy has time to move his reserves; we must take steps to stop him moving them. So we arrive at the *ninth conclusion*, that the primary task of the air striking force in a land battle must be to isolate the area attacked from reinforcement and supply; and thus to ensure that the impetus of the attack on the ground is not checked by enemy reserves rushed to the threatened point by road or rail.

The other point concerns the framing of a plan of battle. Mr. Winston Churchill, in the remarkable memorandum written in 1917, quoted in the first chapter of this book, wrote as follows: 'For our air offensive to attain its full effect it is necessary that our ground offensive should be of a character to throw the greatest possible strain upon the enemy's communications.' Mr. Churchill was ahead of his times in those days and his words, so prescient for the date when they were written, may well serve as the *tenth conclusion* to this review of the influence of air action in land warfare, written nearly twenty years later. It is no longer a matter of the soldier making his plan for battle on the ground and then turning to see how the air can help him. Land and air operations must be deliberately planned to get the best out of each other; and the plan of campaign on the ground, whether in attack or defence, may be profoundly influenced by the air factor. After their offensive on the Aisne in May 1918 had been brought to a standstill north of the Marne, the Germans would have been well advised, under modern conditions, to cut their losses and retire at least behind the Vesle. When the failure of their subsequent attack in July was followed by Pétain's counter-offensive on the 18th, there were thirty German divisions in the huge salient between Compiègne and Rheims; and even if other considerations had not led to a decision to attack in this quarter, here was obviously the place where offensives on the ground and in the air could best play into each other's hands. For the only railways and all the main roads leading into that great salient squeezed through the bottle-neck of Soissons;[1] it is surely not impossible that even in 1918, had the Allies concentrated every available aeroplane in continuous attack on Soissons from the moment when Mangin's tanks left their starting line on the morning of July the 18th, we might have 'pulled tight the string of the bag' and put 'paid' to the remains of those thirty German divisions. Even as it was, Ludendorff, in an Order to the armies dated August the 4th, explained the defeat by saying that 'it was not the enemy's tactical successes which caused our withdrawal, but the precarious state of our rearward communications'—an explanation which, though no doubt only partly true, was probably somewhere near the mark.

Thus, the commander having selected a suitable point of attack, the tasks of the two Services in contributing to his object in a battle on land can be clearly and easily defined. That of the forces on the ground is to turn or, if that is impossible, to penetrate the enemy's defences, defeat his forward troops, and press on to the occupation of the objective. That of the air striking force is to attack the communications serving the enemy's army, to disorganize and delay the reserves coming in to support his forward troops, and to prevent traffic bringing up food, ammunition, and the mass of other material essential to their continued resistance. These two tasks, though distinct, are mutually interdependent. The men on the ground impose upon the enemy a situation in which an intensive flow of reinforcement and supply is vital; while the men in the air, by blocking that flow, create the opportunity for their comrades on the ground to progress unchecked to the point where their action becomes decisive.

[1] Note also how well off the French were relative to the enemy in this sector— seven lines of railway and plentiful roads served the French front from Compiègne to Rheims.

Prophecy is notoriously a hazardous occupation; and this review of air action in land warfare can be little more than a tentative groping after truth. The writer has tried to abstain from the role of advocate, and to confine himself to a reasoned and impartial examination of the case for air power. The march of events may prove him wrong—some new development may enormously enhance the power of the defence in the air, or the efficacy of passive defences from the ground; something as yet quite untried, some ray or some equivalent of the mine-fields of naval warfare, may upset our calculations; at least it may be found that we have in some respects overstated, in others underestimated, the influence of air power on armies. In its wider implications in the political field there seem, on the whole, to be grounds for hope that the terrible potentialities of air action on the cities may, at least in part, be offset by the stabilizing influence of the added strength which it confers upon the defence against invasion on the ground. As to that, it is perhaps too early to hazard a definite opinion. In the more limited sphere to which this book has been confined there is one *general conclusion* of paramount significance, to which all others seem to point. No attitude could be more vain or irritating in its effects than to claim that the next great war—if and when it comes—will be decided in the air, and in the air alone. But it is surely not too much to assert that, for serious warfare against a first-class enemy, the days of National Armies on the traditional, early twentieth-century, man-power and shell-power model are inevitably numbered. The air is only one, but it is the most decisive one, of a number of factors favouring the rise of the small, highly mobile, hard-hitting, armoured and mechanized army of to-morrow. Can it really be supposed that we shall ever see again in the face of air action the millions of men, the thousands of tons of ammunition, the network of trenches stretching half-way across Europe, and the vast organizations at the bases and on the lines of communication that turned northern France into a passable imitation of Epsom Downs on Derby Day?

Here then is a new Army of a Dream.[1] First, the defence forces, the garrisons of permanent frontier fortifications, of anti-aircraft and coast defences—supplemented on mobilization by

[1] With apologies to the late Rudyard Kipling.

volunteers, and in continental armies by the older classes of reservists. Behind them, the counter-attack forces, the mobile armoured formations, backed by motorized infantry and sappers, armed mainly with machine-guns and anti-tank weapons for consolidation and for protection of the armoured force at rest. And in the air above them—the long-range striking force, which from behind the cover of the frontier defences will carry the war into the enemy's country, cut the communications behind his fighting troops, and co-operate with the armoured force in the counter-attack. This may be a fantasy—and in any event the fundamental modifications which it involves in our British system of army organization will be difficult and drastic. But if the conclusions reached in this book be sound, if their implications be fully realized and accepted, and these modifications or something on these lines be found essential—then there will surely not be lacking a modern Cardwell to put them into effect.

APPENDIX A

G.H.Q. Programme for the employment of the R.A.F. in the Battle of Amiens—August 8th, 1918.

Formation	Z−2 day and Z−1 day	Night of Z−1/2	Z day	Night of Z/Z+1	Z+1 day
Second Army	Normal activity	Normal activity	Normal activity	Normal activity	
Fifth Army	Increased activity	Normal activity	Normal activity	Normal activity	
First Army	Increased activity	Normal activity	Normal activity. Some fighter squadrons to be held in readiness to operate on Third Army front if required.	Normal activity	
Third Army	Normal activity	Normal activity	Day-bombing squadron to operate on northern portion of Fourth Army front. Fighter squadrons to be held in readiness to co-operate on Fourth Army front if required.	Night-flying squadron to work on Fourth Army front north of the Somme.	To be decided later in accordance with the course of operations on Z day.
Fourth Army	Normal activity. Arrival of additional squadrons not to be disclosed by additional bombing or patrols.	Normal activity	Fourth Army aviation to be employed in accordance with orders of G.O.C. Fourth Army. No. 6 Squadron to co-operate with Cavalry Corps. No. 8 Squadron to co-operate with tanks.	Night-flying squadron to work on Fourth Army front south of the Somme.	
9th Brigade R.A.F.	Working on First and Fifth Army fronts and resting.	Some Handley-Page machines to fly over the area of tank concentration to drown noise of the tanks.	Bombing attack on enemy aerodromes in early morning. Fighter squadrons to work in conjunction with bombers. Fighter squadrons to be held in readiness subsequently to counter enemy activity on Fourth Army Front. Bombing attack on enemy communications in the evening. Fighter squadrons to work in conjunction with the bombers.	Two night-flying squadrons (one Handley-Page) to continue the attack on hostile communications carried out by day bombers in the evening. A second Handley-Page squadron to attack communications leading to Fourth Army front from the north.	

APPENDIX B

TABLE 1

EMPLOYMENT *of bombing squadrons in the Battle of* AMIENS, *August 8th to* 11*th inclusive, reckoned by individual aircraft attacks against* (*a*) SOMME *bridges and* (*b*) *other objectives.*

Bde.	Sqn. No.	Aircraft on charge	8th (a)	(b)	9th (a)	(b)	10th (a)	(b)	11th (a)	(b)	Percentage of effort expended.
5th	205	20	16	31	42	..	31	..	..	38	100
,,	(N) 101	17	10	..	20	..	19	..	..	..	30
9th	27	19	36	..	21	..	6	..	..	9	50
,,	49	17	29	..	8	..	5	..	6	..	35
,,	98	17	19	..	14	..	..	7	..	7	35
,,	107	18	17	13	21	..	..	4	..	6	40
,,	(N) 83	19	15	..	14	..	14	..	7	..	30
,,	(N) 207	10	..	..	7	..	12	..	6	..	30
,,	(N) 215	10	..	..	..	7	..	7	..	7	25
1st	18	19	..	8	..	10	8	17	..	13	35
10th	103	19	..	7	..	6	10	16	6	5	30
3rd	57	18	..	20	..	10	..	20	..	9	40
,,	(N) 102	18	20	..	27	..	..	22	..	24	60
Totals 13	..	221	162	79	173	33	105	93	25	118	..
			241		206		198		143		

TABLE 2

Analysis of expenditure of effort by No. 205 Squadron standard:
100 per cent. One Squadron (No. 205)
60 ,, One ,,
50 ,, One ,,
40 ,, Two Squadrons
35–25 ,, Eight ,,
Average just under 50 per cent.

APPENDIX C

TABLE I

Actual air concentration on August 8th, 1918. (a) Squadrons (by numbers) and (b) aircraft on charge. By classes

Engaged on AMIENS *front August 8-11th*									*Elsewhere in* FRANCE, *not engaged* AMIENS							
Day bombers		Night bombers		Fighter recce.		Fighters		Bde.	Day bombers		Night bombers		Fighter recce.		Fighters	
(a)	(b)	(a)	(b)	(a)	(b)	(a)	(b)		(a)	(b)	(a)	(b)	(a)	(b)	(a)	(b)
27 49 98 107	19 17 17 18	83 207 215*	19 10 10	62	18	32 73 1 43 54 151	19 26 19 24 23 19	9th Bde. (G.H.Q. Res.)	25	17	58	17				
205	20	101	17	48	21	23 201 24 84 41 80 65 209	19 19 19 19 19 25 25 19	5th Bde. (Fourth Army)								
57	18	102	18	11	18	3 60 56 87	25 19 19 19	3rd Bde. (Third Army)								
18	19							1st Bde. (First Army)			148	18	22†	17	19 40 64 203 208	19 18 19 19 26
								2nd Bde. (Fifth Army)	206	17	149	18	20	18	29 70 74 79 85	19 24 20 18 19
								10th Bde. (Second Army)	108 211	18 18			88	18	17Am 148Am 2 Aust 4 ,, 46 92	19 18 19 25 27 19
103	19							5th Gp. (V.A.D.P.)	202 217 218	15 14 17	38 214	18 10			204 213 210	24 24 19
								Ind. Force	55 99 104	18 15 18	100 216 97§	18 11 10				
Totals 8	147	5	74	3‡	57	18	376		10	167	8‖	121	3	53	19	395

* Transferred to Independent Force August 19th.
† This squadron did one escort task on the SOMME front between the 8th and 11th.
‡ Of these squadrons numbers 62 and 11 and two flights of 48 were actually used entirely as fighters
§ This squadron joined the Independent Force on August 9th from Home Establishment.
‖ Two additional Handley-Page squadrons (Nos. 110 and 115) joined the Independent Force from Home Establishment on August 31st.

TABLE 2

Suggested amendments to meet plan suggested (see p. 188)

Add to Table 1 concentration for AMIENS				Bde.	To remain in other sectors			
Day bombers	Night bombers	Fighter recce.	Fighters		Day bombers	Night bombers	Fighter recce.	Fighters
(a) (b)	(a) (b)	(a) (b)	(a) (b)		(a) (b)	(a) (b)	(a) (b)	(a) (b)
25 17	58 18			9th Bde.				
	148 18		19 19 60 18 64 19	1st Bde.			22 17	203 19 208 26
206 17	149 18 .		29 19 70 24 74 20	2nd Bde.			20 18	79 18 85 19
108 18 211 18			2 Aust 19 4 ,, 25 46 27 92 19	10th Bde.			18 18	17Am 19 148Am 18
202 15 217 14	38 18 214 10		204 24 213 24	5th Gp.	218 17			210 19
55 18	100 18 97 10 216 11			Ind. Force	99 15 104 18			

Amend Table 1. Totals

15 264	13 195	3 57	30 633		3 50		3 53	7 138

APPENDIX D

TABLE 1. *Distribution and nominal Strength of German Aircraft on the WESTERN Front on March 21st and August 8th, 1918.* (Note. *Actual Strength can be taken as about two-thirds of these figures.*)

Number of squadrons by classes—figures in brackets represent establishment of aircraft per unit (see note on p. 184).

Army (Sea to SWITZERLAND)	Recce. (6) 21/3	(6) 8/8	Reinforced Recce. (9) 21/3	(9) 8/8	Photo R. (9) 21/3	8/8	Serio Phot. (4) 21/3	8/8	Battle (6) 21/3	8/8	Fighter (14) 21/3	8/8	Bomber (6) 21/3	8/8	Total Aircraft in army 21/3	8/8
4	10	13	..	..	..	..	1	1	1	2	4	5	6	6	162	200
6	6	10	..	1	..	..	1	1	..	2	4	5	..	3	96	173
17	11	5	6	3	..	..	..	..	7	2	13	6	3	..	362	153
2	8	7	8	5	..	..	1	..	11	2	10	6	3	3	348	201
18	8	6	7	5	..	1	1	1	9	1	12	4	6	3	382	165
Total *aircraft* on the front of attack, i.e. in 17, 2, and 18 armies on 21 March '18	162	..	189	..	9		8	..	162	..	490	..	72	..	1,092	..
9	..	7	..	2	..	..	..	..	..	7	4	8	..	..	..	214
7	5	12	4	10	..	..	..	2	..	12	2	16	3	3	140	484
1	5	10	..	7	..	..	..	..	..	8	3	14	..	3	58	385
3	11	4	..	..	..	..	..	..	..	3	4	4	..	3	84	116
5	3	4	..	..	..	..	..	..	..	..	2	1	..	..	122	38
Detmt. C	4	5	..	..	..	..	..	..	..	..	2	4	..	..	46	44
19	4	4	..	..	..	..	2	1	..	..	3	2	3	3	70	98
Detmt. A	4	4	..	..	..	..	..	..	..	..	3	2	..	..	74	56
Detmt. B	7	5	..	..	..	..	..	..	..	..	5	2	..	..	112	58
Total units	89	96	25	33	1		6	6	28	39	68	78	24	27	..	..
Total aircraft	534	576	225	297	9		24	24	168	234	952	1,092	144	162	approx. 2,056	2,385

Notes. (a) See also Appendix XVI to *The War in the Air*, vol. iv. (b) The above figures exclude three squadrons and two marine fighter units on the BELGIAN coast. This omission does not vitiate the proportions shown in Table 2, because it is offset by the fact that the British naval co-operation units stationed the other side of Dover Straits are not included in the British figures.

TABLE 2. *Totals in Bomber and Fighter (including 'Battle') aircraft, British and German, on March 21st and August 8th. Note especially when comparing British and German figures that the German whole front was from the sea to* SWITZERLAND, *while the British represented less than half the Forces opposed to the Germans on that front.*

		Nominal strength, i.e. establishment	Approximate actual strength	Remarks
21st March, 1918:				*Germans attacking*
German strength. Whole Front	1	1,264	842	
" " On front of attack	2	724	482	Approx. 57 per cent. of 1 (in 17th, 2nd, and 18th Armies)
August 8th, 1918:				*British attacking*
British strength. British front	3	..	1,390	British attack complete surprise, German air units concentrated against French in CHAMPAGNE, and behind LYS front in north.
" " On front of attack	4	..	654	Aircraft serviceable: i.e. in Third and Fourth Armies, and 9th Bde. R.A.F., 47 per cent. of 3.

When the existence of the French Air Force is taken into account (about 2,700 aircraft on March 21st, 1918), these figures make it clear that the *proportion* of their total available strength concentrated at the decisive point of attack by the Germans was probably at least twice the proportion of *their* total strength concentrated by the Allies in similar circumstances. Actually it was 10 per cent. more than the British did. It speaks very well for the strategical handling of the German air force, and it is as well for us that their tactical performance did not come up to the same high standard.

APPENDIX E

GERMAN *divisions identified on Fourth Army front August 8th to 11th inclusive*

(Authority: *The Story of the Fourth Army, Appendix D*)

A. *In line opposite IVth Army 8.viii.18 North to South*	B. *In reserve IInd German Army 8.viii.18*	C. *Identified 8–11.viii.18 exclusive of A, with locations on 8.viii.18*
233 Identified 9th off front of attack on 8th. 27 (Wurt.). 108 13 41 117 225 14 Bav. 192	243 RIBECOURT. 54 ROISEL. 43 Res. FLAUCOURT. 109 BERNY. 107 E. of LE CATELET, not identified August. 21 E. of LE CATELET, identified 13.8.18.	*August 8th.* 54 43 relieving 108. 109 „ 117. *August 9th.* 243 1 Res. from line Eighteenth Army. 82 Res. Eighteenth Army, GUISCARD. 10 Seventh Army, LAON. 119 Sixth Army, TOURNAI area. *August 10th.* 5 Bav. Seventeenth Army, MARQUION. 79 Res. Fourth Army, ROUBAIX area. 121 Fourth Army, COURTRAI area. 221 Seventeenth Army, DENAIN. *August 11th.* Alpine Corps, Fourth Army, GHENT. 26 (Wurt.) Res. Seventeenth Army, FLERS. 38 Sixth Army, LILLE area. 204 (Wurt.) Eighteenth Army in line NOYON.

Order of battle, German Armies from North to South: Fourth Army.
Sixth Army.
Seventeenth Army.
Second Army.
Eighteenth Army.
Ninth Army.
Seventh Army.
First Army.

LINES OF COMMUNICATION BEHIND THE BRITISH SECOND ARMY IN
FLANDERS, MAY 1918

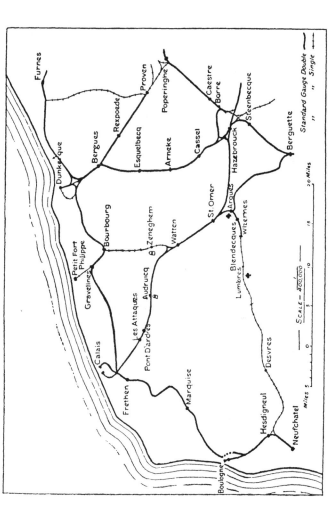

ᗷ Ammunition Depots ; ✦ Supply Depots

Reprinted from THE ARMY QUARTERLY *by permission*

SKETCH-MAP No. 1

LENS

DROCOURT

DO

OPPY

LA SCARPE

ST POL.

Source
de la
Scarpe

ARRAS

SCARPE R.

185
Dury

5-B
M

SEVENTEENTH ARMY

DOULLENS

BAPAUME

ANCRE

THIRD
ARMY

26 W.R

Aveluy

Fiers

Equancourt

SE

Etricourt

Fins

BRITISH

ALBERT

Dernancourt

Méricourt

Moislains

Morlancourt

Oervo

R. SOMME

Bray

Fuilleres

54

R

Etinehem

Cappy

CANAL DU N

Eclusier

PERONNE

SOMME R

Flaucourt

AMIENS

Villers-Bretonneux

OUTER AMIENS

Foucaucourt

poingt

43

Bouvincourt

Estrées-en-Chaussé

FOURTH
ARMY

Hangard

Estrees

Berny

109

Bris

Verm

D.P

LUCE R.

Harbonnieres

LUCE R.

Libons

St Christ-Briest

Athies

Chaulnes

Falvy

Rosieres-en-Santierre

MOREUIL

AVRE R.

Bethencourt

Hangest

Plessier

Arvillers

Damery

Voyennes

Offoy

NESLE

HAM

P

84.R

Pierrepont

AVRE R.

S.R.

Esmery Hallon

ROYE

82.R.

APPROX. FRONT LINE AUG. 8TH

MONTDIDIER

Guiscard

FRENCH

FIRST ARMY

NOYON

LEGEND: ARMY BOUNDARIES   GERMAN DIVISIONS IN RESERVE   RAILWAY   ROAD

FIT   UNFIT

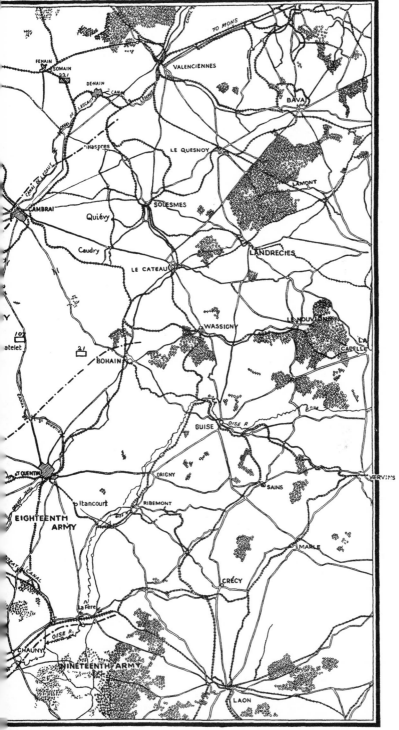

THE EASTERN FRONT, 1914-15

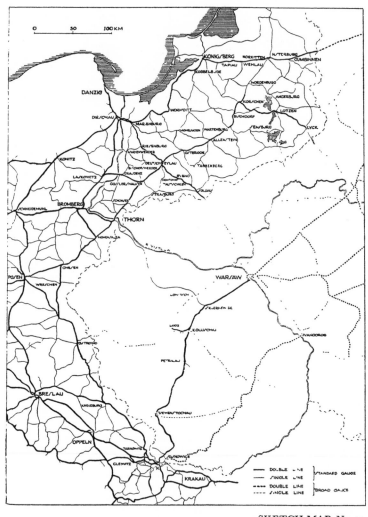

SKETCH-MAP No. 3

INDEX

Aachen Gap, 208.

'Absolute block', system of, 138.

Aerodromes, enemy, attack on, 37; blockade of, examples at Douai and Ramleh, 44; proportion of bombs dropped on, by Independent Force, 53; tendency to underrate possible effects of attack on, 53; British experience in France, 54 et seq.; possibility of bomb-proof accommodation at, 54; dangers of hangars in war, 55; aircraft must be widely dispersed on, 56; concentration of many aircraft on single, to be avoided, French experience at Verdun, 57; when widely dispersed, efficient meteorological service and intercommunication essential, 57; attack on, facilities for replacement of casualties in relation to, 59; attack on, time factor in relation to, 60; orders for bombing of, at 3rd Ypres, 127; enemy, bombed on morning of August 8th, 156, 172; number of, in Amiens sector, March 1918, 54, 189. *Specific:* Bertangles, 54; Bushire, 56; Coudekerque, 55; Detling, 71; Dunkirk, 58; Gontrode, 24; Imbros, 145, 202; Luxeuil, 71; Marquise, 58; Ochey, 71; Ramleh, 44; St. Denis Westrem, 24; Varssenaere, 186.

Africa, South, deficiencies of British Army in, made good by 1914, x.

Aim, *see* Object.

Aircraft, effect of better, as factor in air superiority, Palestine and France, 1917, 27; numbers of in France on Aug. 8th, 183, 188, and App. C; *see also* Bombers, Fighter and Assault aircraft.

Air fighting, relation of, to air plan as a whole, example in Independent Force, 7; strategical conditions of, much the same as of land warfare, 32; in battle of Amiens, 154–60.

Air forces, enemy, reasons for air action against, 3; attack on, in relation to object of air force, 7; independent of lines of communication, 8; not committed to any one course, 9; enemy, attack on, conditions which may render it necessary to devote main effort to, 27; action against, in relation to plan as a whole, method of approach to problem of, 28–30; action against, a joint responsibility of fighters and bombers, 32; theory that they can only be brought to action by consent, 32; can only be dealt with effectively by offensive, 34; not a battle-field weapon, 90; cannot hold, therefore must have room to make their weight felt, 93; Foch's directive to Allied, 105, 127–8, 194; German, in battle of Amiens, 160–2.

Air Staff, at R.A.F. H.Q. in France, 77, 128; memoranda on bombing of railways, 125–6, 169; orders for 3rd Ypres, 127; on night bombing of railways, 135.

Air superiority, true status of, a means to an end, 4, 5, 10; official definition of, 4; object of, 4; analogy with naval term, 4; not a definite stage to be gone through, 10, 27; with one exception, 39; in Palestine campaign, 11–13; in Somme battle, 13–14; two main principles, the main offensive and supplementary offensive, 15–16; subject to variation, according to factors, 27; at Cambrai, 99; at battle of Amiens, 163, 181.

Aisne, German attack on, May 1918, 72, 212.

Allenby, demands up-to-date squadrons in Palestine, 11; use of air power in Palestine, 12.

Amiens, bombing of rail triangle at, 132 n.; congestion at, during Somme battle, 133; battle of, on August 8th, 1918, detailed survey of, 148–99.

Ammunition, policy for supply of, in 1914, x; expenditure of, at Messines, 112, 118; depots destroyed at Campagne, 118; Audruicq, 119; Spincourt, 119; trains hit by bombs at Valenciennes, 125; St. Quentin, 94, 125; Thionville, 125, 132, 134; Lillers, 125 n.; supply to IIIrd Australian Corps by air on Aug. 8th, 153, 156.

Appreciation, of employment of air force at battle of Amiens, 176 et seq.; of situation on enemy railways with view to bombing, by H.Q. Second Army, 115; in connexion with Passchendaele battle, 127.

Armoured forces, less vulnerable than other troops, 92; attacks by, without

Germany, first bombing of, as reprisal, 65, 71; early attempts at bombing of, 70; Trenchard's views on bombing of, 71, 75; bombing of, an unjustifiable diversion in April 1918, 78.

Gontrode, enemy aerodrome at, 24.

Groups, *see* Brigades.

Groves, Brig.-Gen. P. R. C., on 'Independent' air operations, 69; Director of Flying Operations at Air Ministry, 1918, 75.

Gunpowder, the first revolution, 200.

Guynemeyer, Capitaine Georges, 27.

Haig (F.-M. Lord), 148, 149, 150; has to send two fighter squadrons to England, summer 1918, 24.

Hangard, bridge over R. Luce at, 145, 173.

Hangars, dangers of, in war, unnecessary, 55, 190.

Hareira, Turks caught by assault aircraft in Wadi, 90.

Head-quarters, effects of bombing of, general, 97; Liman von Sanders's at Nazareth, 97; larger head-quarters ideal air objective, 97; central air H.Q. proposed for battle of Amiens, 196.

Hoeppner, Gen. von, assumes command German air force, 1917, 27 n.; quotations from, 20, 190 n.

Hohenzollern bridge at Koln, 145, 146.

Holborn, effects of gas explosion in, 143.

Home Defence Force, composition of, in 1918, 187.

Horse, day of, in serious warfare gone for ever, 94; vulnerable to air action, 95; numbers of, in horsed units, 95; *see also* Mounted troops.

Imbros, operations by British aircraft during Gallipoli campaign from, 145, 202.

Independent Force, 2; relation between air fighting and true object of, 7; tonnage of bombs dropped by, 53; action of, possible effects in causing collapse of Germany, 65; origin and growth of, in 1918, 70–8; first proposed to War Cabinet, 73; Inter-Allied Force first proposed, 74; decision on, 74; negotiations at G.H.Q. and R.A.F. H.Q. in France, during formation of, 76–8; unjustifiable diversion in April 1918, 78; mobility disregarded, 78; in July 1918, 148; composition of, Aug. 8th, 186, and App. C.

Indian frontier, importance of roads as communications on, 143.

Industrial areas, *see* Production.

Intelligence, need for, in choosing objectives for diversions, 25; combined Inter-Service Bureau, need for, 26; air defence, warning system, need for in the field, 52; use of, in making the air plan, 88, 122; in connexion with railway bombing, 133, 140; in battle of Amiens, information available, 170, 176.

Interior lines, attacker always on, in the air, 23.

Isolation of the battle-field, the object of air force in battle, 167, 177, 212.

Italian front, rout of Austrian armies by air action on, 103.

Japanese, war *v.* Russia 1904, 201; possibilities of modern campaign against Russia in Manchukuo, 203.

Junctions, as objectives, 129, 132–4; at Amiens, congestion during Somme battle, 133; need for detailed intelligence of, 134; at Deutsche Eylau and Bromberg, density of traffic at, 133; at Metz Sablon, blocked by Independent Force, 134; at Thionville, destroyed by Independent Force, 134; Muslimieh, on Palestine railway, 205.

Kantara, single railway line of supply into Palestine from, 203.

Khyber, an army operating beyond the, 202.

Köln (Cologne), District office, and area controlled by, 137; Hohenzollern bridge at, 145; rail traffic across Rhine at, 146.

Könitz, collision at, effects of, 123.

Kosturino pass, rout of Bulgarian army by air action in, 102.

Kresna pass, rout of Bulgarian army by air action in, 102.

Kuleli Burgas, bridge over R. Maritza at, 145.

Lagny, French railway disaster at, 138.

Land campaign, definition of, 2.

Liége, effect of demolitions in 1914 on rail bottle-neck at, 135, 209.

Lihons, capture of, by Australian Corps, Aug. 11th, 159.

Lillers, 125 n.

Limburg Appendix, 208.

Locomotives, as air targets in attack on railways, 140.

Longeau, German bombing of railway triangle at, 132 n.